T0093897

The Physics of Planet Earth and Its Natural Wonders

Dmitry Livanov

The Physics of Planet Earth and Its Natural Wonders

 Springer

Dmitry Livanov
National University of Science
and Technology MISIS
Moscow, Russia

ISBN 978-3-031-33425-2   ISBN 978-3-031-33426-9   (eBook)
https://doi.org/10.1007/978-3-031-33426-9

This Springer imprint is published by the registered company Springer Nature Switzerland AG
The registered company address is: Gewerbestrasse 11, 6330 Cham, Switzerland

Preface

The book you are holding in your hands is about the physics of the world. The world around us, which is made up of the planets and stars that we see when looking at the sky, mountains and rivers on the Earth's surface, the seas and oceans with their storms and lulls, our planet's atmosphere with lightning and thunder, wind, snow and rain—all of this is like a huge laboratory in which physical experiments are taking place every minute and every second. Steven Hawking wrote, "The subject of science is often taught in school in a dry and boring way. Children learn to mechanically memorize material in order to pass tests, but do not see any connection between science and the world around them." The aim of this book is to show that this connection really does exist and to explain several physical phenomena, which we encounter every day.

Over the centuries people have asked the question: Why is our world like this? By providing us with the knowledge that we need, physics gives us information about the world, an understanding of what happens in nature and why and also predicts what will occur in the future.

What is the special significance of physics for the development of our civilization and what distinguishes it from other natural sciences?

First, while describing and explaining natural phenomena, physics constructs a scientific picture of the world of modern man. Everyone should have at least a general idea of how the world in which they live works. This is fundamental not only for our general development; a love for nature implies that we also respect everything that happens in it. In order for this to happen, we need to understand the laws that cause these natural processes to take place so that we leave our children a world in which they can live. Neither all properties of the material world nor all laws of nature have been studied; nature is still fraught with many mysteries. As physics develops, we become more knowledgeable about the world around us.

Second, physics determines mankind's technological development. Everything that distinguishes modern-day civilization from society of past centuries has arisen as a result of practical application thanks to discoveries in physics. Research in the field of electromagnetics, for example, led to the development of household electrical appliances, cell phones and the Internet, which are so essential today, while discoveries in mechanics and thermodynamics resulted in the production of automobiles and trains. Moreover, advancements in the physics of semiconductors gave rise to the unveiling of the computer, while in aerodynamics, airplanes, helicopters and rockets were developed. In return, innovations in engineering and technology make it possible to conduct fundamentally new research.

Third, physics forms the foundation of all the other natural sciences—astronomy, chemistry, geology, biology and geography—because it explores fundamental common factors. Chemistry, for example, studies atoms and molecules, the substances of which they are composed and the transformation of one kind of matter into another. The chemical properties of a substance are determined by the physical properties of atoms and molecules, which are described in such branches of physics as thermodynamics, electromagnetics and quantum mechanics.

Fourth, physics is closely connected with math because math provides a framework with which the laws of physics can be precisely developed. Physical theories are almost always formulated as mathematical equations. Mathematical formulas had to be included in this book, as they make the essence of physical phenomena clearer. Math makes it possible to quantify what occurs around us and to establish common factors and connections between physical quantities, thus making it possible not only to explain, but also to predict, and, in so doing, take control over the future. Without question, only those mathematical relationships that can be verified and measured observationally and with experiments are of value in physics. Furthermore, the level of complexity of mathematical tools should correspond to the approximation of the physical model that is used. Everyone knows the joke made by Albert Einstein, Nobel Laureate in Physics, who, when referring to using overly complex mathematical tools in physics said, "Since the mathematicians have invaded the theory of relativity, I do not understand it myself anymore." Therefore, the level of mathematical description used for each problem in this book is of the simplest nature and does not go beyond the scope of the material that is presented in a school curriculum. It is also usually limited to qualitative explanations and approximate estimates.

Fifth, observations and experiments form the basis of physical research. By generalizing them, it is possible to highlight those patterns that are overarching and the most substantial, as well as aspects of observed phenomena. In the early stages of experiments, these underlying characteristics are primarily empirical, i.e., they describe only the properties of physical objects and not the internal operations that produce these properties. By analyzing empirical regularities, physicists use appropriate mathematical tools to develop physical theories, which explain the phenomena being researched based on today's ideas of the structure of matter and the interaction between its constituent parts. In so doing, this gives clarity to the way that systems work and reasons for the occurrence of different phenomena. General physical theories help to formulate the laws of physics, which are undisputed until large quantities of new experimental results do not require that they be clarified and reviewed.

I invite you to venture into the fascinating and complicated world of physics. I will end this short preface with a quote from another Nobel Laureate in Physics, Peter Kapitza, who said, "Nothing prevents a person from becoming smarter tomorrow than they were yesterday."

Prof. Dmitry Livanov
Doctor of Physical
and Mathematical Sciences
National University of Science
and Technology MISIS
Moscow, Russia

Acknowledgements Author would like to thank Dr. Jill A. Neaendorf for the excellent work of translating the book from Russian into English. The English translation of the book would not have been possible without the extensive support and help of my friend and colleague Dr. Timothy E. O'Connor.

About This Book

This book is meant for high school students, university students, professors and teachers of physics, as well as everyone who wants to understand what is happening in the world around them and develop a scientific perspective on the vast number of natural phenomena that exist. Every section of this book has essentially a set of physics problems, which enable the reader to strengthen their understanding of physical laws and learn to apply them in interesting situations.

Contents

About the Author

Dmitry Livanov has held a variety of roles in his life. He began his career in science in the field of theoretical physics and will be remembered for several widely publicized articles in the field of superconductivity and the physics of metals for which he was awarded the Golden Medal of the Russian Academy of Sciences for Young Scientists. After developing his own course on solid-state physics and writing his own textbook, Dmitry became heavily involved in teaching. Within the following decade, Dmitry climbed the ranks from Associate Professor to Chancellor of the Moscow State Institute of Steel and Alloys (MISiS) and during that time he did a great deal to make MISiS into a modern-day European scientific and technical university. While holding the position of Minister of Education and Science of the Russian Federation, he worked from 2012 to 2016 on reforming Russia's system of science and education. However, regardless of the capacity in which he works or the position he holds, Dmitry remains, first and foremost, a scientist with creative drive and a rational perspective on the world.

1

The Earth in the Solar System

Abstract We start with a discussion of the two milestones of Nature—the law of universal gravitation and Kepler's laws, and the latter is the sequence of the first. These laws account for the formation of the Solar system. The Sun is considered with special attention as the main source of energy inside the Solar system. Then we review the main physical features of the planets in the Solar system. The rotation of the Earth around its axis is then discussed, and the associated physical phenomena on the Earth's surface as well. In concluding the first chapter, we look at the physical background of our calendar.

Planet Earth is the home of all human beings and people have long sought to understand how it works. What is the shape of our planet? Why and how does it move in relation to the Sun and stars? Why do different phenomena on the Earth's surface, deep inside of it and around it occur exactly as we see them? These are perhaps the primary questions that mankind has always sought to answer.

To our ancient ancestors, the Earth seemed to have a flat surface like that of a disk resting on elephants or turtles (Fig. 1.1). They reasoned that a starry sky, through which heavenly bodies moved, hung above the flat Earth. Today such an idea would make even elementary students laugh, but at that time it was an excellent concept. It explained all natural phenomena: the Earth seemed flat to someone standing on it and earthquakes were thought to be caused by the movement of that very gigantic animal supporting the Earth's

© The Author(s), under exclusive license to Springer Nature
Switzerland AG 2023
D. Livanov, *The Physics of Planet Earth and Its Natural Wonders*,
https://doi.org/10.1007/978-3-031-33426-9_1

Fig. 1.1 The Earth as imagined by our ancient ancestors

foundations on its back. No one had seen the ends of the Earth because it is so big. Moreover, our ancestors understood the concept of "down" as a direction perpendicular to the Earth as a disk.

However, more than two thousand years ago, the ancient Greeks understood that the Earth is round. Aristotle, the great philosopher of ancient times, was the first to prove that the Earth has a spherical shape. He noticed that during a lunar eclipse, the shadow of the Earth is round and the constellations that are visible from the Earth change places when one travels along its surface. Aristotle surmised that the motionless Earth was located in the center of the world, around which all cosmic bodies rotate in circular orbits (Fig. 1.2). This was called a *geocentric model*. Today we sometimes think in terms of the geocentric system when we say, for example, that "the sun rises" and we imagine that it emerges from a motionless forest instead of a forest that is rotating around the Earth's axis. However, every child today knows that the Earth revolves around the Sun in its orbit (Fig. 1.3). Moreover, the Earth, just like a top, spins around its axis. But what is the shape of the orbit of the Earth and of other planets? Does the angle between the plane of the Earth's orbit and the axis of its rotation change? And why don't the Earth and other planets fly away from the Sun, and the Moon fly away from the Earth? What are their laws of motion? We will examine these questions in the first chapter of this book with the help of physics and math.

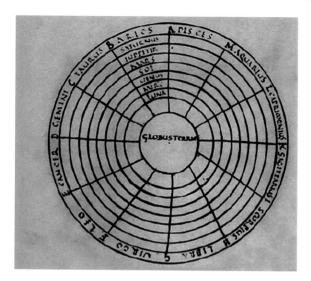

Fig. 1.2 One of the earliest images of the geocentric system that has survived. Macrobius, a manuscript from the ninth century BC

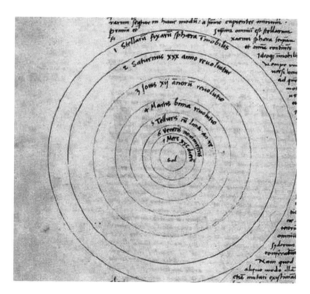

Fig. 1.3 The Solar system. An illustration from Nicolaus Copernicus' book *On the Revolutions of the Celestial Spheres*, 1543

1.1   The Law of Universal Gravitation and Kepler's Laws

Astrologists of the Middle Ages unsuccessfully tried to predict the life events of specific people based on the movement of celestial bodies. Their predictions turned out (and still do) quite badly, but their observations and descriptions of the movement of planets have been extremely useful. The Dane Tycho Brahe, who developed new methods of observation that made it possible to minimize measurement errors and achieved a level of accuracy that was unprecedented for the sixteenth century, made particularly correct predictions. Thanks to data from his observations, Johannes Kepler discovered the laws of planetary motion in the seventeenth century. Based on these laws, Isaac Newton formulated the law of universal gravitation in his book *The Mathematical Principles of Natural Philosophy*, which was published in 1687.

> Newton introduced the law of universal gravitation in his book *The Mathematical Principles of Natural Philosophy*, which was published in 1687, and in which he did not mention anything about the gravitational constant. It was only after a little more than 100 years, in 1798, that Henry Cavendish introduced it on an experimental basis and the formula took on its final form.

This is how the historical chain of discoveries progresses, but the logic of physical theories does not always coincide with this progression. Although Kepler's laws were discovered prior to the discovery of the law of universal gravitation, we will first consider this law as the reason for the movement of celestial bodies, and thereafter examine Kepler's laws as a result of the law of universal gravitation.

The law of universal gravitation quite simply states:

> Two material points with masses m_1 and m_2 are mutually attracted and the force from their mutual attraction is proportional to the product of their masses and inversely proportional to the square of the distance between them.

$$F = G\frac{m_1 m_2}{r^2} \tag{1.1}$$

If we are not dealing with material points, but with round bodies of finite sizes, then the law of universal gravitation will include the distance between the centers of these spheres (Fig. 1.4). However, not only round bodies are attracted, but also bodies of any shape. In the latter case, it is necessary to split each of the bodies into very small parts and sum up the interaction of these parts in order to get the force of gravitational pull.

Now it becomes clear that the direction "down" coincides with the direction of the force that acts on a body by the Earth. In this case, "down" is the direction toward the center of the Earth.

The law of universal gravitation is one of the most important laws of physics. It is both simple and universal. From atoms and molecules to stars and galaxies, this law is applicable to all bodies of the universe, the distance between which is much larger than their size. But why don't we notice, for example, a pull between books lying on a table? The reason is because of the *G* coefficient, which is called the *gravitational constant*. Its value is very small: $G = 6.67 \times 10^{-11} \frac{m^2}{kg\,s^2}$.

Consequently, the force of gravitational pull becomes noticeable only when a body's mass is not just large but very large! Which body close to us has the largest mass? The Earth, of course. This is precisely why we feel the pull of all bodies toward the Earth, which we call gravity, and we absolutely cannot detect any gravitational pull of objects on a table toward each other.

Thus, according to the law of universal gravitation, let us assume that a planet and the Sun are pulled toward one another: force is directed along a straight line that connects the centers of their mass and is inversely proportional to the square of the distance between them. If these conditions are met, the movement of bodies is described by Kepler's three laws.

Kepler's First Law The path of the planets around the Sun is elliptical in shape, with the center of the Sun being located at one focus.

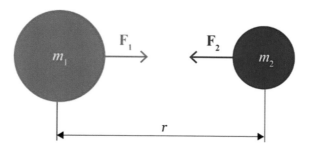

Fig. 1.4 The forces of gravitational pull acting between two bodies

Planets do not move around the Sun in a perfect circle, as ancient astronomers believed, but rather in elliptical orbits (Fig. 1.5). The scientific community was not eager to accept this fact because by default it was thought that "the celestial sphere was the epitome of perfection" and the circle was officially considered the most perfect geometrical figure. Therefore, all celestial bodies were "required" to move only in a circle. However, to this end nature had its own opinion, which had to be reckoned with.

An Ellipse What exactly is an ellipse? Figuratively speaking, an ellipse is an elongated circle (this definition is sometimes seen in crossword puzzles). A circle has a center and the distance between the center and any point of the circle is the radius, which is always the same. Now imagine that there are two centers and they have begun to separate. In order to imagine this, we will conduct a small experiment. We will take a piece of paper (cardboard is better), poke two needles or pins into it that are about 5 cm (1.97 in.) apart, connect them with a ring of thread and then, while pulling on the thread with a pencil, draw a line, making sure that we are always pulling on the thread (Fig. 1.6). Now we have an ellipse!

Half of the "length" of the ellipse is called the *semi-major axis* and is denoted by *a*, and half of the "width" of the ellipse is called the *semi-minor axis* and is denoted by *b*. If we move the needles further apart from each other, the ellipse will be more elongated; if we move them closer together, it will be less elongated. In the most extreme case, when the needles are very

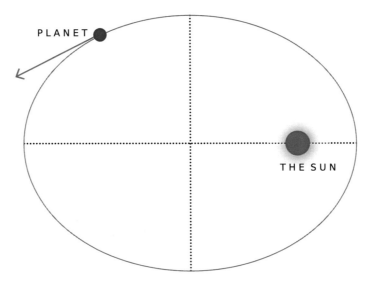

Fig. 1.5 A planet's orbit in the Solar system

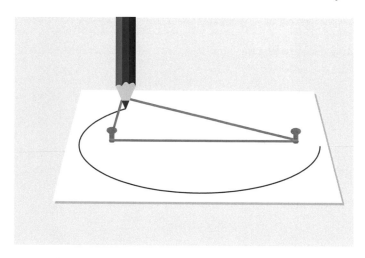

Fig. 1.6 How to draw an ellipse

close together, the "width" is equal to the "length" and a circle forms, i.e., $a = b$. The elongation of the orbit of a celestial body is determined by the eccentricity $e = \sqrt{1 - \frac{b^2}{a^2}}$ ($e = 0$ is a perfect circle and $e = 1$ is when the ellipse degenerates into a line segment).

The equation of an ellipse is:

$$\frac{x^2}{a^2} + \frac{y^2}{b^2} = 1 \tag{1.2}$$

If $a = b$, the equation of an ellipse turns into a center-radius form with a radius of a.

An ellipse is usually characterized by the value of the semi-major axis a and of the eccentricity $e = \sqrt{1 - \frac{b^2}{a^2}}$. The foci of the ellipse are two points that are symmetrically located on a large axis and the distance between them is equal to $2ae$ (Fig. 1.7). Those who are interested in geometry can easily prove that for any ellipse point, the sum total of the distances to the foci is constant and equal to $2a$.

We will calculate the area of an ellipse. In order to do this, we imagine a cylinder with a height h, a radius of the base b, and a volume equal to $V = \pi b^2 h$. We cut the cylinder along a plane at an angle α (Fig. 1.8a). An ellipse with semiaxes $a = \frac{b}{\cos a}$ and b is obtained in the cross section.

We attach the truncated top of the cylinder to it from below (Fig. 1.8b), but the volume of the cylinder does not change. Now we cut the cylinder into a large number of n disks that are parallel to the new base. The area of

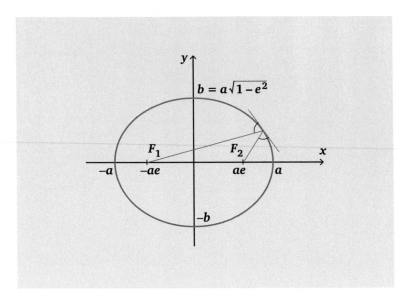

Fig. 1.7 Parameters of an ellipse

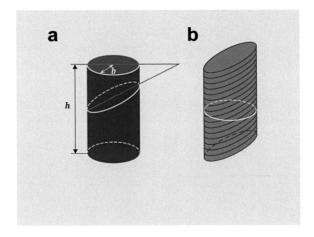

Fig. 1.8 How to calculate the area of an ellipse

each disk is S and the height is $\frac{h}{n}\cos\alpha$. When we make the volume of the cylinders equal, we get $S = \pi ab$.

Now that we know what an ellipse is, we can move on to Kepler's second law.

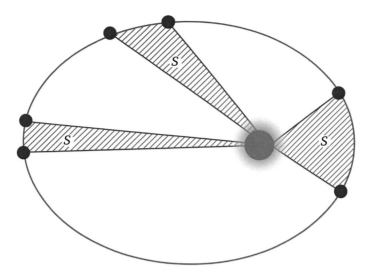

Fig. 1.9 Illustration of Kepler's second law

Kepler's Second Law The radius vector drawn from the Sun to each planet sweeps out equal areas in equal intervals of time (Fig. 1.9).

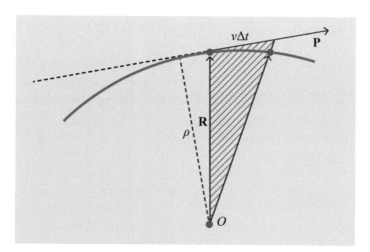

Fig. 1.10 Illustration of Kepler's second law

Support for Kepler's Second Law Let us consider a planet with the mass m moving in the field of gravity of the Sun, which is located at point O. We will disregard the influence of other celestial bodies on the planet's movement.

We will denote the planet's speed as **v**. Its momentum is then $\mathbf{P} = m\mathbf{v}$ and is directed along the tangent to the planet's trajectory (Fig. 1.10).

We will fix the origin of coordinates at point O and drop the perpendicular from it onto a line that is defined by the vector **P**. We will denote the length of that perpendicular as ρ. The product is called the *angular momentum*:

$$L = \rho P = \rho m v. \tag{1.3}$$

Because the moment of gravity relative to the origin of coordinates is zero, the angular momentum of the planet relative to the Sun does not change when the planet moves. During Δt the planet orbits the distance $v\Delta t$. Let's consider a shaded triangle with the base $v\Delta t$ (see Fig. 1.10). Its area is

$$\Delta S = \frac{1}{2} v \Delta t \rho = \frac{1}{2} \frac{L \Delta t}{m}. \tag{1.4}$$

If Δt time is short, then the base of the triangle practically coincides with the portion of the trajectory through which the planet passes. In this case, the triangle itself is a section of the area that the radius-vector **R** of the planet sweeps out during Δt. Since the angular momentum is constant in time, the area swept by the line segment is proportional to the time interval Δt, that is, for equal periods of time the radius-vector of the planet will sweep out equal areas. This is the principle of Kepler's second law.

Let's imagine that an imaginary thread connects the Sun and a planet. The area over which the planet has passed remains constant each time for the same intervals of time. By applying Kepler's second law, we can easily calculate the linear speed of the planet, the velocity value of which can greatly differ depending on the place where the planet is located at that particular moment. In perihelion, which is the point in a planet's orbit that is closest to the Sun, planet speed is at its maximum, while in aphelion, which is the furthest point from the Sun, planet speed is minimal. Therefore, the speed of the Sun has the highest possible velocity value in perihelion $v_{max} = 30.3$ km (18.83 mi)/s. In the furthest point in orbit the formula is $v_{min} = 29.3$ km (18.21 mi)/s. This is why in January when the Earth reaches its perihelion, the Sun's speed in the sky is a little bit faster than in July when it is at aphelion. Admittedly, it is very difficult to observe this with the naked eye due to the fact that the shape of the Earth's orbit is almost circular.

However, this is not the case with other celestial bodies such as comets. Many of them travel on extremely elongated paths. For example, the orbital eccentricity of Halley's Comet is 0.967. Imagine that you are flying on that comet further and further away from the Sun until it is nothing more than a bright star. Your speed in relation to its speed becomes slower and slower... In the darkness and silence, you travel for decades toward aphelion by cosmic standards at a snail's pace of 0.9 km (0.56 mi)/s. Now you have passed through aphelion and the comet starts to pick up speed. The Sun keeps expanding and finally the comet passes through perihelion with incredible speed—54.5 km (33.86 mi)/s! During that very short trip radiation from the Sun causes the comet's surface to quickly become very hot. Owing to this, particles of the comet frantically evaporate and it grows a tail millions of kilometers (hundreds of miles) long. Imagine if the Earth had the same eccentricity as Haley's Comet. Without any sunshine in aphelion the temperature would drop to almost zero; even the air would freeze and precipitation would fall on the cold and lifeless surface. In perihelion the Sun would turn into a brutal fiery ball that would make oceans dry up and rocks melt.

Kepler's Third Law The ratio of the squares of the orbital periods of planets around the Sun (Fig. 1.11) is equal to the ratio of the cube of the length of the semi-major axis of its

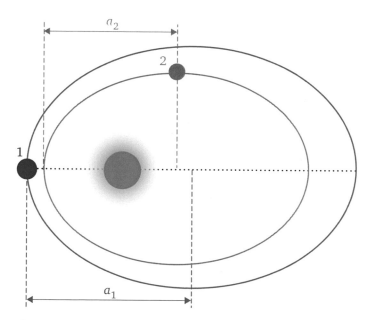

Fig. 1.11 Illustration of Kepler's third law

elliptical orbit:

$$\frac{T_1^2}{a_1^3} = \frac{T_2^2}{a_2^3}.$$

(1.5)

Connection Between Kepler's Third Law and the Law of Universal Gravitation Newton deduced the law of universal gravitation from Kepler's third law. Let's try to follow his train of thought.

Let's assume that there are several planets that move around a star and for simplicity's sake, let's say that this movement follows a circular pattern. We will denote the radii of planetary orbits as R_1, R_2, etc., and their orbital periods around the star as T_1, T_2, etc. Based on Kepler's third law it follows that:

$$\frac{R_1^3}{T_1^2} = \frac{R_2^3}{T_2^2} = \cdots = \text{const.}$$

(1.6)

After introducing the angular velocity of the planets $\omega = \frac{2\pi}{T}$, Eq. 1.6 can be written as follows: $\omega_1^2 R_1^3 = \omega_2^2 R_2^3 = \cdots = \text{const.}$

Newton assumed that the force of interaction of a planet with a star is an exponential function of the distance between them, i.e., it follows that: $F = AR^n$.

Then the accelerated velocity that the planet receives when it comes in contact with a star is proportional to the distance as well: $a = BR^n$.

Newton understood that when movement occurs around the periphery of a circle, centripetal acceleration is proportional to the squared velocity and inversely proportional to the distance, i.e.,

$$a = \frac{v2}{R} = \overset{2}{\omega} R \Rightarrow \omega^2 = BR^{n-1}.$$

(1.7)

By virtue of Kepler's third law, the product $\omega^2 R^3$ should have a constant value, specifically, it does not depend on distance. On the other hand, $\omega^2 R^3 = BR^{n+2}$, which shows that the required condition is met in the case of $n = -2$. In that event, $\omega^2 R^3 = B$. Newton also surmised that the constant value B is proportional to a star's mass M: $B = GM$. For acceleration we then get: $a = G\frac{M}{R^2}$.

The force that passes such acceleration on to the planet with the mass m will be equal to:

$$F = G\frac{Mm}{R^2}.$$

(1.8)

This is the law of universal gravitation. Newton is not to be commended as much for the fact that he discovered a way to express the force of gravitational pull as he is for universalizing this law.

The derivation of Kepler's third law is also quite intriguing.

We will find the time of a planet's complete period of rotation around the Sun, which is the orbital period T. According to Kepler's second law, during Δt the radius-vector of the planet sweeps out the area $S = \Delta t \frac{L}{2m}$. This means that one can calculate the orbital period after having divided the area of the ellipse by the sweep speed: $T = \frac{S}{\frac{L}{2m}}$.

The area of the ellipse is equal to $S = \pi a b$ where b is its semi-minor axis. Then $T = \frac{2\pi a b m}{L}$.

From the laws of conservation of energy and momentum, we can obtain the formula $b = \frac{L}{\sqrt{2mE}}$ and then $T = \pi a \sqrt{\frac{2m}{E}}$. After expressing energy in terms of the semi-major axis $a \left(E = \frac{GmM}{2a} \right)$, we get:

$$T = \frac{2\pi}{\sqrt{GM}} a^{\frac{3}{2}}. \tag{1.9}$$

What is the use of knowing about Kepler's third law? First, we can compare planets' orbits. Second, when we understand the orbital period of a celestial body, we can find the point of the semi-major axis of its orbit. Alternatively, after measuring the point of the semi-major axis of a celestial body's orbit, we can confidently determine what its orbital period is. The further a celestial body is from the Sun, the longer its orbital period.

Kepler's laws have proven just how versatile they are. In particular, they "work" well not only when calculating the orbits of celestial bodies around the Sun, but also when determining the parameters of the motion of man-made satellites and natural satellites of other planets. Information obtained from studying other galaxies has validated that Kepler's laws are carried out in outer space, which makes it possible to receive a great deal of significant and fascinating data.

It was recently reported that astronomers discovered a galaxy in which Kepler's third law does not "work": in this particular galaxy, which has a high velocity of rotation, hydrogen should have been emitted into more distant orbits, but this did not happen. This is because in this galaxy mass is "in short supply." But this is precisely how the natural sciences differ from the liberal arts in that laws that have been discovered and proven cannot be "incorrect" or "out of date." If you find out that a law that has proven its validity millions of times over can suddenly be disproven, that can only mean one thing— there is some new factor at work here that is unknown to you. This was the case in this situation. If we assume that in a galaxy there is a significant mass of a certain type of matter that we have not observed, then a law will once again be applied. In order to find this additional mass, scientists estimated the mass of gas between the stars. But that was not enough. Modern-day physics was faced with a mystery—what is the invisible substance called "dark matter"? Kepler, who had formulated his laws several centuries ago, still contributes to the development of modern-day science.

1.2   A Star Called the Sun

The most important place in the homes of ancient people was the hearth. It gave off warmth, light and it was where people cooked. In those days when people did not know how to make a fire, a cold hearth could lead to the death of an entire tribe. The Sun, just like the hearth, gives light and warmth to the entire Solar System. Without the Sun no life forms on the Earth could exist. This is exemplified by the fact that in our energy-based biosphere there are planets that store the Sun's energy in the process of photosynthesis. It is not a surprise that in the religions of different countries the Sun God (Ra, Helios, or Jarilo) always existed and was among the most highly revered and powerful gods (Fig. 1.12). Therefore, we will devote some attention to our "cosmic hearth," under the rays of which life began and exists.

The mass of the Sun is $M_S = 1.99 \times 10^{30}$ kg. Although it is difficult for us to imagine such a weight value within the categorical concepts to which we are accustomed, when speaking about celestial bodies such a quantity is nothing out of the ordinary. The Sun's mass makes up no less than 99.9% of the entire Solar System. In a manner of speaking, the mass of the Sun is the mass of the entire Solar System. Therefore, the Sun's supremacy over all other celestial bodies in the Solar System cannot be doubted. Its mass is large enough to keep planets and other celestial bodies of the Solar System in orbit around it.

But on the scale of the Galaxy, the Sun is of average size and an ordinary star; there are, according to various estimates, between 200 and 400 billion such planets in our Galaxy alone, which is called the *Milky Way*. We are located deep in the Galaxy (Fig. 1.13) at a distance of 26,400 light years from the center of it. It is a quiet place and it is precisely there, according to one hypothesis, that the speed of the stars and the spiral arms of the Galaxy come together. For this reason, it is difficult for us to fall out of the Galaxy, overtake our neighbors or lag behind them. This is called the *corotation circle* and we are very lucky to be located inside of it. After all, if a collision occurs between a celestial body in our Galaxy and another star, the existence of something as insignificant as our planet will not be of great concern to our celestial neighbors. However, thanks to the corotation circle we have little reason to worry about this actually happening.

Fig. 1.12 The Sun God Ra, ancient Egypt, 901–713 BC

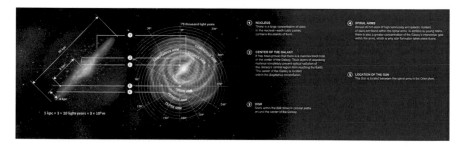

Fig. 1.13 The structure of our Galaxy

The galactic year of our planet, i.e., one complete revolution around the center of the Galaxy, is approximately 200 million years.

Since we have already described the Sun's location, now we need to speak about its age. The Sun was formed approximately 4.5 billion years ago when a molecular cloud composed of hydrogen, helium and other elements rapidly compressed under the influence of gravitational force. A star with a mass such as that of the Sun has a lifespan of approximately 10 billion years. Thus, according to the standards of a star, the Sun is in its prime.

Ancient astronomers already knew the average distance from the Earth to the Sun: $R_{E-S} = 1.496 \times 10^{11}$ km, which stems from the laws of gravitational astronomy. If one were to fly on an airplane at the speed of 800 km (497 mi)/h, it would take more than five years to cover this distance. However, it takes a beam of light 8 min and 19 s to do this.

Because of the fact that from the Earth the Sun resembles a ball with an average angular diameter $\alpha_S = 9.3 \times 10^{-3}$ rad $= 31' 59''$, it is easy to calculate the Sun's radius. It is: $R_S = 6.7 \times 10^8$ m, which is 109 times greater than the Earth's radius. The average solar density is $\rho_S = 1.4 \times 10^3$ kg/m^3. We see that solar density is just slightly greater than water density and approximately four times less than the average density of the Earth.

The internal structure of the Sun is well studied today. With the help of various devices, including spectroscopes and different types of telescopes, the Sun's electromagnetic radiation is recorded in a variety of ranges and its surface and activity are observed, which enables us to draw conclusions about its internal structure.

The chemical composition of the Sun mostly consists of hydrogen (about 90%) and helium (about 9%) atoms. The remaining elements (iron, oxygen, nickel, nitrogen, silicon, sulfur, carbon, magnesium, neon, chromium, calcium and sodium) account for less than 2%.

The primary value of the physical characteristics both on the surface and in the inner regions of the Sun, as well as the nature of energy that the Sun (and other stars) constantly emits, is also well known. How did this information become known? After all, it is impossible to fly to the Sun and measure its temperature with a thermometer. Knowing physics and mathematics help make everything clearer.

Constitutive Relation of Solar Matter We will identify some of the physical characteristics of the processes that take place on the Sun.

The intensity of solar radiation is characterized by a value called the *solar constant*. It is the total solar radiation energy per unit of area perpendicular to the Sun's rays and at the Earth's average distance from the Sun. According to data obtained from exo-atmospheric measurements, the solar constant is: $S = 1367$ W/m^2. Despite its name, the solar constant does not remain constant over time. Its value is determined by two main factors: the distance between the Earth and the Sun, which changes throughout the year (the annual variation is 6.9%, which is from 1412 W/m^2 at the beginning of January to 1312 W/m^2 at the beginning of July) and changes in solar activity.

If we multiply the solar constant by the area of the sphere with a radius R_{E-S}, we can find out how much total energy is emitted by the Sun in 1 s, i.e., the solar radiation output or the solar luminosity. It is equal to $L = S \times 4\pi R_{E-S}^2 = 3.83 \times 1026\,\text{W}$. We can also find the solar energy flux density, i.e., the amount of energy that is emitted per second by a square meter (square foot) of the Sun's surface. This is, in fact, its brightness: $R = \frac{L}{4\pi R_S^2} = 6.29 \times 10^7\,\text{W}$. The energy flux density emitted by an object is related to its temperature according to the Stefan–Boltzmann law: $R = \sigma_B T^4$, where $\sigma_B = 5.67 \times 10^{-8}\,\frac{\text{W}}{\text{m}^2\,\text{K}^4}$. From here it follows that we can calculate the temperature on the Sun's surface: $T_{S.surf} = 5780\,\text{K}$.

We can obtain the constitutive relation of solar matter, specifically, the relationship between temperature, pressure and density. If the gas on the Sun's surface consists mainly of electrically neutral atoms (weakly ionized plasma), then when immersed deep into the Sun and when the temperature and pressure are increased, the electrons of the atomic shells detach themselves from their atoms, thus forming plasma, the degree of ionization of which reaches 100%.

Let's assume that solar plasma is made up of hydrogen nuclei (protons), helium nuclei and electrons in a ratio of 91:9:109, respectively. Solar plasma is, in fact, a mixture of three gases: hydrogen nuclei, helium nuclei and electrons of the same temperature. For each of the gas mixtures, which can be considered ideal, the constitutive relation $p = nk_B T$ is true, where the concentration of gas particles is $n = \frac{\rho}{m}$ (ρ is the gas density and m is the mass of gas particles) and $k_B = 1.38 \times 10^{-23}\,\text{J/K}$.

In addition, the total plasma pressure is the sum total of the pressure of individual gases: $p = p_H + p_{He} + p_e$, while the overall concentration is $n = n_H + n_{He} + n_e$. We will designate the fraction of hydrogen ions in the total number of particles as N. Then the fraction of helium ions will be proportional to $1 - N$, and the fraction of electrons will be proportional to $2 - N$. The concentration, pressure and gas density of hydrogen ions can be written as

$$n_H = \frac{N}{3 - N}n, \tag{1.10}$$

$$p_H = \frac{N}{3 - N}nk_B T, \tag{1.11}$$

$$\rho_H = m_H \frac{N}{3 - N}n. \tag{1.12}$$

This is similar for the gas of helium ions:

$$n_{He} = \frac{1 - N}{3 - N}n, \tag{1.13}$$

$$p_{He} = \frac{1 - N}{3 - N}nk_B T, \tag{1.14}$$

$$\rho_{He} = m_{He}\frac{1 - N}{3 - N}n. \tag{1.15}$$

For the gas of electrons, it is:

$$n_e = \frac{2 - N}{3 - N}n, \tag{1.16}$$

$$p_e = \frac{2 - N}{3 - N}nk_B T. \tag{1.17}$$

$$\rho_e = m_e \frac{2-N}{3-N} n. \tag{1.18}$$

We will take into account that the mass of a helium ion is four times greater than the mass of a hydrogen ion, $m_{He} = 4m_H$, while the mass of the electron is small to negligible as compared to the mass of the proton, $m_e \ll m_H$. Then we can assume that $\rho_e \approx 0$. We get:

$$\rho = \rho_H + \rho_{He} + \rho_e = \frac{4-3N}{3-N} m_H n. \tag{1.19}$$

The constitutive relation of solar matter will take on a modern-day appearance of the ideal gas law (also known as the Mendeleev-Clapeyron equation):

$$\rho = \frac{\rho}{\mu m_H} k_B T, \tag{1.20}$$

for gas with a molar mass:

$$\mu = \frac{4-3N}{3-N} = 0.61. \tag{1.21}$$

This mass turned out to be very small due to the fact that, although electron gas exerts pressure, it has no bearing on a change in density. This explains the low density of solar plasma.

Since we know the constitutive relation, we can find the temperature and pressure in the central region of the Sun. The pressure, which is created inside the Sun, is due to the gravitational compression of matter. If we consider a column of matter with the density p and the height H in a gravitational field with an acceleration of gravity g, the pressure it creates will be equal to: $p = \rho g H$. This formula can be roughly used in this case, although the rate of acceleration of gravity for stars naturally varies with depth. We get:

$$p_S \approx \rho_S g_S R_S \approx \frac{GM_S^2}{R_S^4} \approx 10^{15}\,\text{N/m}^2. \tag{1.22}$$

By using an equation of condition, one can even estimate the temperature of the central region of the Sun: $T_S \approx \frac{p_S m_H}{k_B \rho_S} \approx \frac{GM_S m_H}{k_B R_S} \approx 2 \times 10^7\,\text{K}$.

Our estimate of 20 MK roughly corresponds to information about precise calculations. But is gravitational energy enough for the Sun and other stars to exist? We will estimate the potential energy of the Sun after it has been compressed by the force of gravity: $E_p \approx \frac{GM_S^2}{R_S} \approx 4 \times 10^{41}\,\text{J}$.

This energy can provide the Sun's brightness that we see: $L = 3.83 \times 10^{26}\,\text{W}$ for a period of $t = \frac{E_p}{L} \approx 3 \times 10^7$ years.

The lifetime of the Sun can, in fact, last almost as long as five billion years. The solution to our above-calculated equation illustrates that, in addition to gravitational energy, a different and much more powerful energy source is needed to warm the Sun and other stars.

We will first focus on the structure of the Sun (Fig. 1.14) and the primary physical features of its layers (Fig. 1.15). Thereafter, we will analyze those physical mechanisms that cause physics to be at work in the universe.

The central part of the Sun is the *core* having a radius of approximately 151,000 km (93,827 mi). Matter in the core is extremely dense. It is about 1.5×10^5 kg/m^3, which is 150 times higher than water density. The temperature in the center of the solar core exceeds 1.5×10^7 K.

It seems to us that the Sun is a burning ball, but burning is actually an example of a chemical change, while an energetically more powerful process occurs in the solar core: a thermonuclear reaction that makes hydrogen nuclei fuse into helium nuclei. Every second the Sun loses 4.3 T (4.74 sh. tn.) of hydrogen. But there is no need to worry—scientists estimate that because the Sun has a mass of 2×10^{27} T, there is enough solar fuel to last about five billion years.

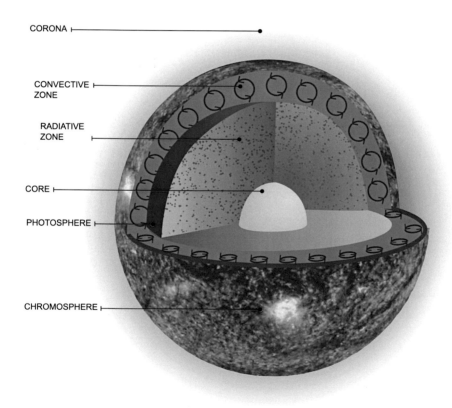

Fig. 1.14 The structure of the Sun

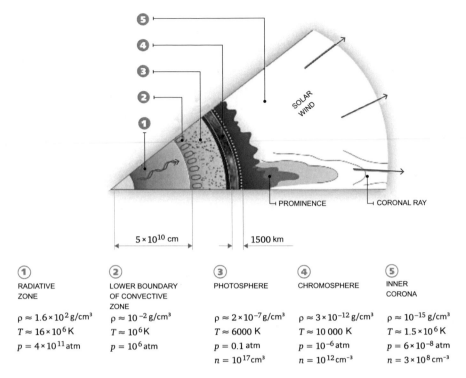

①	②	③	④	⑤
RADIATIVE ZONE	LOWER BOUNDARY OF CONVECTIVE ZONE	PHOTOSPHERE	CHROMOSPHERE	INNER CORONA
$\rho \approx 1.6 \times 10^2\,\text{g/cm}^3$	$\rho \approx 10^{-2}\,\text{g/cm}^3$	$\rho \approx 2 \times 10^{-7}\,\text{g/cm}^3$	$\rho \approx 3 \times 10^{-12}\,\text{g/cm}^3$	$\rho \approx 10^{-15}\,\text{g/cm}^3$
$T \approx 16 \times 10^6\,\text{K}$	$T \approx 10^6\,\text{K}$	$T \approx 6000\,\text{K}$	$T \approx 10\,000\,\text{K}$	$T \approx 1.5 \times 10^6\,\text{K}$
$p = 4 \times 10^{11}\,\text{atm}$	$p = 10^6\,\text{atm}$	$p = 0.1\,\text{atm}$	$p = 10^{-6}\,\text{atm}$	$p = 6 \times 10^{-8}\,\text{atm}$
		$n = 10^{17}\,\text{cm}^3$	$n = 10^{12}\,\text{cm}^{-3}$	$n = 3 \times 10^8\,\text{cm}^{-3}$

Fig. 1.15 The physical features of the layers of the Sun

Energy of the Sun Let's estimate how much energy is released during thermonuclear reactions and the amount of solar matter consumed by them per unit of time. Understanding the nature of the Sun and other stars' inner energy sources forms the basis of Einstein's 1905 discovery of the theory of special relativity. If a system's mass changes during a reaction, then this change Δm is counterbalanced by a change in the energy of the system according to the formula:

$$\Delta E = -\Delta mc^2, \tag{1.23}$$

where c is the speed of light. It follows that when any transformation of matter occurs, the sum $E + mc^2$ remains the invariant. When chemical reactions occur, they are followed by the release or absorption of energy; a relative change in mass is negligibly small and usually cannot be experimentally detected. If elementary particles are involved in the reaction (this type of reaction is known as a *nuclear reaction*), then a relative change in mass is noticeable. The gravitational pull that groups protons and neutrons together in atomic nuclei functions at short distances—around 10^{-15} m (33^{-49} ft)—but it is very strong. Thus, any realignment of the atomic nucleus requires a great deal of work in order to resist this force. If a change in the mass of the system occurs during a nuclear reaction, energy is released.

This energy, which is the kinetic energy of particles and photons that are flying in different directions, as well as of quantum radiation, ultimately transforms into heat, which vigorously heats up a substance.

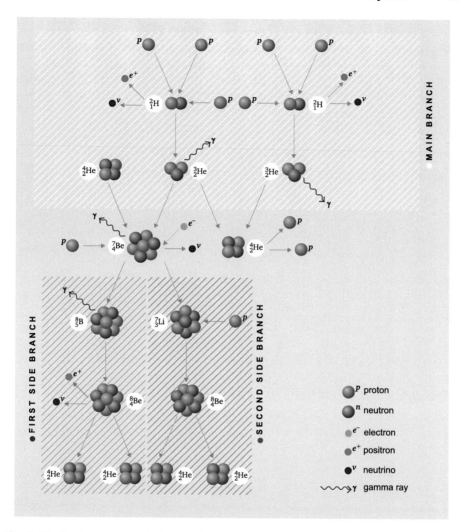

Fig. 1.16 Proton-proton chain reaction

When conducting possible nuclear reactions, one must remember that the interaction of particles is realized only when certain conservation laws are satisfied during these reactions. There are three of such conservation laws. The first is the law of conservation of charge, according to which during a reaction the full electric charge should not change. The second conservation law calls for conserving what is referred to as the *baryon number*, which is the total number of protons and neutrons before and after the interaction. Finally, the third law of conservation dictates lepton number conversation. Electrons, muons and taumesons (and corresponding neutrinos) are called *leptons* and conserving a lepton number means that when any interactions occur, the difference in the number of leptons and their antiparticles remains constant.

Since the largest number of hydrogen nuclei (protons) is located in solar plasma, we will first consider the reaction that occurs when two protons collide. If there are two protons on the left side of the reaction, a deuteron—which is formed from a nucleus of the heavy isotope of hydrogen known as *deuterium*—must be on the right side in order to fulfill the law of conservation of a baryon number. In order to conserve an electric charge, a positively charged particle, i.e., a positron, must be added to the right side. But if a positron, that is, an antiparticle of the election, appears on the right side, then it is necessary to add a neutrino to the right side in order to converse a lepton charge. Thus, the resulting reaction appears as follows:

$$^1p + {}^1p \rightarrow {}^2d + e^+ + v \qquad (1.24)$$

The positron that is created will dematerialize with one of the many electrons:

$$e^+ + e \rightarrow 2\gamma. \qquad (1.25)$$

The result is:

$$^1p + {}^1p + e \rightarrow {}^2d + v. \qquad (1.26)$$

Now let's determine the mass balance of the reaction. The particle mass that has reacted (the left side) is equal to 2.01575 AMU (atomic mass units). Since a neutrino does not have any mass, the mass of the reactor products (the right side) is 2.0142 AMU. Therefore, the mass defect, i.e., the loss in particle mass during the reaction, is $\Delta m = 0.00155$ AMU. The amount of energy $E = \Delta mc^2 = 2.3 \times 10^{-13}$ J released during the reaction largely changes into a neutrino's kinetic energy. What happens next? When a helium isotope is created, a deuteron will react with a proton:

$$^2d + {}^1p \rightarrow {}^3\text{He}. \qquad (1.27)$$

In this reaction, $\Delta m = 0.0059$ AMU and the energy emitted is $E = \Delta mc^2 = 8.8 \times 10^{-13}$ J. Since neutrinos are not generated during the reaction, this energy is entirely used to heat the surrounding solar matter. The isotope of the helium atom ^3He is relatively stable. However, when exposed to high temperatures and pressure, this isotope combines with one just like it to produce the ^4He isotope and two protons:

$$^3\text{He} + {}^3\text{He} \rightarrow {}^4\text{He} + {}^1p + {}^1p. \qquad (1.28)$$

For this reaction, $\Delta m = 0.0138$ AMU and $E = \Delta mc^2 = 2.1 \times 10^{-12}$ J. The nuclei of the ^4He isotope do not react any further. The latter reaction occurs in 65% of cases. In the remaining 35%, the following reactions take place:

$$^3\text{He} + {}^4\text{He} \rightarrow {}^7\text{Be} + v; \qquad (1.29)$$

$$^7\text{Be} + e \rightarrow {}^7\text{Li} + v; \qquad (1.30)$$

$$^7\text{Be} + {}^1 p \to {}^8\text{B};\qquad\qquad (1.31)$$

$$^8\text{B} \to {}^8\text{Be} + e^+ + v.\qquad\qquad (1.32)$$

A diagram of the proton-proton chain reaction is shown in Fig. 1.16.

In either case, the proton-proton (PP) chain ends with the creation of the nucleus of a helium isotope ^4He from four protons. In addition to protons, two electrons reacted. The products of the cycle, in addition to the helium isotope, are two neutrinos. The total mass defect of the proton-proton chain is equal to 0.0287 AMU. Thus, the formation of one helium nucleus is followed by a release of energy that is equal to 4.3×10^{-12} J. If we roughly consider that a neutrino takes all of the energy that is released because of the first reaction of the cycle, which is about 4.5×10^{-13} J, then we see that as a result of each proton cycle, approximately 4×10^{-12} J of energy is spent on heating solar matter.

Interestingly, the first reaction of the proton-proton chain (the interaction of two protons) is the slowest, which is believed to limit the rate at which energy is released in a star and ultimately the star's evolution over time. The reason for this is that since two protons have the same charge, the probability of their interaction is very small. Let's estimate how often the protons of solar plasma react. We already know that the luminosity (the full power of radiation) of the Sun is $L = 3.8 \times 10^{26}$ W and as a result of each thermonuclear reaction of a helium isotope , 4×10^{-12} J of energy is released. After dividing one by the other, we can determine how many helium atoms form on the Sun in 1 s, which is 10^{38}. But how many total hydrogen atoms are there in the Sun? Once we have divided the Sun's mass by the mass of a hydrogen atom, we get the answer—10^{57}. Hence, every second about a 10^{-19}th part of solar matter is used in nuclear reactions. To put this a different way, out of 10^{19} protons only one pair interacts each second. Our Universe is ten billion years old, which means that during its entire lifetime less than 10% of stellar matter will have time to be consumed. This means that the thermonuclear source of solar energy is virtually inexhaustible.

Thermonuclear reactions occur only in the Sun's core. How is the tremendous amount of energy that is generated by the nucleus transferred to the outer layers? As is well known, there are three types of heat transfer: thermal conductivity; convection and radiation. Let's figure out which type is most appropriate in this case. Since there is a great deal of pressure near the core, convection cannot occur because it requires that layers of matter be mixed together. Energy transfer due to thermal conductivity is also difficult. The reason for this is simple: in order for the process of heat transfer to occur, it is necessary to have a rapid change in temperature. Heat transfer through thermal conductivity, for example, will take place when one end of an ash rake is in the flame of a fireplace and the other is in a basin of ice water. But if the ash rake is completely thrust into a large bonfire, heat transfer will not occur because both ends of it will be heated to the same temperature.

The only type of heat transfer that is left is radiation. A particle of electromagnetic energy known as a photon, which is hydrogen, immediately sets off at the speed of light. But a photon's journey is short because after having moved only 1 μm (one millionth part of a meter) (10^{-5} in.), it is absorbed by a nuclear atom. The nucleus center then heats up and reradiates a photon of the same wavelength. The reradiated photon penetrates the next micron before being absorbed by another nucleus and the process repeats itself. Since a photon does not have any preference in which direction it goes, it moves randomly and often in the opposite direction from the outer surface of the Sun toward its core. This means that photons "roam" around in this area for an average of 170,000 years! That is why this fascinating area is called the *radiative zone*. The temperature in this zone varies from 7 MK in depth to 2 MK on the surface, while its mass density varies from 20 g/cm^3 (in depth) to 0.2 g/cm^3 (on the surface) of the density of water.

Closer to the surface of the Sun, the mass density decreases, which allows plasma to be stirred up. Therefore, the transfer of energy to the surface of the Sun takes place mainly by the movement of matter itself. This is similar to how water moves in a pan on the stove: when water heats up on a burner, it expands, is pushed up, cools down and settles down. This type of energy transfer is called *convection* and the subsurface layer of the Sun, which has a thickness of approximately 2×10^5 km and is called the *convective zone,* is where this process takes place.

Because of convection, so-called *granules* (Fig. 1.17) appear on the surface of the Sun. Figuratively speaking, granules are like "cooking pots," in each of which the heated plasma rises in the center, gives off energy and settles back down to the depth along the "walls." Granules have a lifespan of 10–15 min, which is approximately the time it takes for solar matter to make a complete revolution around a granule. As one gets closer to the Sun's surface, the temperature drops to 5800 K and the mass density becomes less than 10^{-5} of the density of the Earth's air.

But now our long-suffering photon, which was picked up by a convection stream, has finally burst out onto the Sun's surface, which is known as the photosphere. The *photosphere* is gas. According to different estimates, the height of its column reaches 100–400 km (62–249 mi). Most of the Sun's optical radiation comes from the photosphere. As the temperature gets closer to the outer edge of the photosphere, it drops to about 4400 K. In such conditions as these, hydrogen stays primarily in a neutral state. When we look at the Sun, we see the photosphere.

The next outer layer of the Sun that surrounds the photosphere is called the *chromosphere*. It has a thickness of about 2×10^3 km. Solar matter called *spicules* is constantly ejected from the upper boundary of the chromosphere. An average of 60,000–70,000 spicules are observed at a time. The temperature of the chromosphere increases with altitude from 4000 to 20,000 K. The

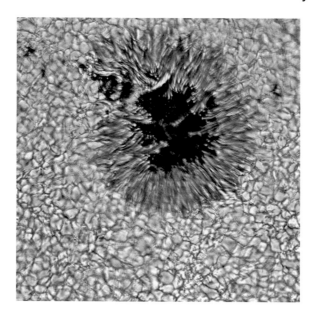

Fig. 1.17 Granulation around a sunspot

chromosphere can be observed only when there is a total solar eclipse and the Moon blocks the light from the brighter photosphere. It is at this moment that the chromosphere becomes visible in all of its glory (Fig. 1.18).

Fig. 1.18 The chromosphere of the Sun during a total solar eclipse (a pink ring around the Moon is visible)

The *corona*, which is the last outer layer, crowns the Sun. Gas emissions similar to large arches called *prominences* (Fig. 1.19) form the basis of the Sun's corona. Within two to three months, prominences are able to stretch out their "tentacles" more than 50,000 km (31,069 mi). Upon reaching this altitude, they may erupt and transmit large amounts of matter into space at speeds up to 1000 km (621 mi)/s. These eruptions can last from several minutes to several hours. When emitted matter reaches the Earth, a geomagnetic storm occurs followed not only by the beautiful northern lights (Fig. 1.20), but also by unpleasant phenomena such as disrupted radio communication and malfunctioning radio electronic equipment. Additionally, people who are sensitive to changes in the weather do not feel well.

"The Carrington Event," which was the most powerful geomagnetic storm in history, occurred on September 1, 1859. On that day the largest coronal mass ejection was observed and after 18 h all telegraph systems throughout Europe and North America stopped working. Storms of this intensity occur approximately once every 500 years. If it had happened today, the consequences

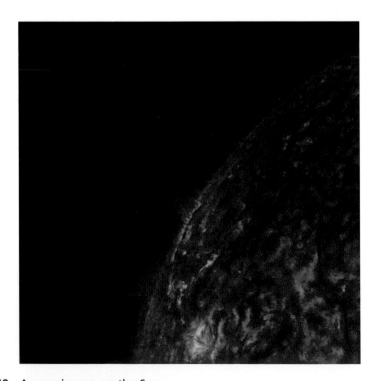

Fig. 1.19 A prominence on the Sun

Fig. 1.20 The northern lights

would have been far more severe than in 1859 because today there is much more equipment, which is essential not just for communication but for our lives in general, that is capable of malfunctioning.

The Sun's corona gives life to solar wind, which is a stream of particles of solar matter. On average solar wind blows about 10^{36} ionized particles into space every second. It is easy to estimate that because of solar wind, the Sun loses about 2.5×10^{-14} of its mass within a year.

Hence, our Sun is a large, complex thermonuclear explosion that has been erupting for more than one billion years. What will happen to it in the future? After all, the existence of our civilization directly depends on it.

Physicists were able to simulate the Sun's future "life journey." As time passes, it will become brighter and brighter. In 1.1 billion years, the Sun will be 11% brighter than it is now and its surface temperature will reach its maximum. This increase in brightness of solar radiation will pose a serious risk to all life forms on the Earth. The Sun's temperature will then gradually drop because of an increase in its volume. Although it will slowly cool down, its brightness will continue to intensify.

In another 2.4 billion years, when the Sun is eight billion years old, it will be 40% brighter than it is now. Because of this, water on the Earth will completely evaporate and the surface conditions on our planet will become similar to those on Venus today.

When the Sun is 10.9 billion years old and the hydrogen that is in its core runs out, it will be twice as bright as it is now. A thermonuclear reaction will occur outside of the nucleus that will spread out further and further into areas far from the Sun's center. It will then begin to grow larger and double in size. The temperature of its photosphere will drop to 4900 K.

After the Sun is approximately 12.2 billion years old, it will become more than 200 times larger and its surface temperature will drop to 2650 K. As a result of a thermonuclear reaction of helium nuclei, the Sun will become thousands of times brighter. Given these conditions, the planets closest to the Sun, including the Earth, will most likely cease to exist.

After helium has been depleted, the Sun will turn into a very hot and very dense star about the size of our Earth. At the beginning of this phase, the Sun's surface temperature will exceed 120,000 K and its brightness will be 3500 times greater than it is now. For billions of years the Sun will then cool down. A "life trajectory" such as this is typical for stars that have a mass like that of the Sun.

It is obviously sad to realize that this planet will someday disappear, but there are several reasons why it is premature for us to worry today that the Sun's future will put mankind in danger. First, there are other threats to our civilization, which makes it quite possible that life on the Earth will disappear before any solar cataclysmic events take place (this, however, is not a very reassuring argument). Second, since the Sun will still be in existence for a long time, if humans do not disappear first, we will definitely come up with some way to save ourselves. Many optimistic science fiction writers think that it is not the Sun that threatens humanity, but, on the contrary, a rapidly developing civilization searching for sources of energy can simply "devour" the Sun.

1.3   The Planets in the Solar System

According to the opinions of today's astrophysicists, the Solar System was formed due to an evolution of a rotating nebula, which originally consisted of solid dust particles and gases. After the particles collided, they became enlarged and the largest number of collisions took place in the center of the nebula where the Sun formed. The terrestrial planets—Mercury, Venus, Earth and Mars—emerged out of large particles of matter that were located some distance from the center of the future Solar System. However, the giant planets—Jupiter, Saturn, Uranus and Neptune (Fig. 1.21)—were formed from light gases scattered on the periphery of the nebula.

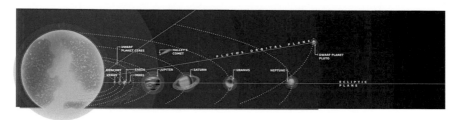

Fig. 1.21 The Solar system. The planets' conjunction is shown

In the twentieth century it was thought that the Sun had nine planets, but in August 2006 at the 26th General Assembly of the International Astronomical Union (IAU) Pluto was "demoted" to the ranks of a dwarf planet. It is curious that the planes of the planets' orbits in the Solar System are almost congruent except for that of Pluto, the ecliptic plane of which is slightly inclined more than 17°.

We will first present the main characteristics of the planets' orbits in the Solar System. What is more, we will begin, of course, with our home planet, whereas the characteristics of other planets will be compared with those of the terrestrial ones.

The Earth's mass is $M_E = 5.976 \times 10^{24}$ kg. The parameters of the Earth's orbit when it revolves around the Sun are as follows: the semi-major axis of orbit is $a = 1.4959787 \times 10^{11}$ m, eccentricity e is 0.017 and the orbital period is $T = 3.1558150 \times 10^7$ s.

The main characteristics of other planets are given in Table 1.1.

Jupiter, which is sometimes called a "failed star," has a very impressive mass. Saturn is distinguishable by its extraordinary rings. One year on Neptune is equal to 164 years on the Earth. One year for us is almost four Mercurian years. Venus and Mars most resemble the Earth and are its nearest neighbors.

Table 1.1 Main characteristics of planets in the Solar System

Planet	Distance to the Sun (in R_{E-S})	Orbital period around the Sun, year	Mass (in M_E)	Eccentricity
Mercury	0.39	0.24	0.056	0.206
Venus	0.72	0.62	0.81	0.007
Mars	1.52	1.88	0.11	0.093
Jupiter	5.20	11.87	3200	0.048
Saturn	9.54	29.46	95	0.056
Uranus	19.18	84.01	15	0.047
Neptune	30.06	164.82	17	0.009

Today there is a lot of information about the planets. It is extremely fascinating to mentally travel to each planet and imagine what the landscape there looks like. I hope that if you have not already done this you will definitely do it in the future. But we are interested in something else—how do these planets affect the Earth?

As one might easily guess, Jupiter, by virtue of its mass, and Venus, because of its close proximity to the Earth, exert the greatest influence on our planet. But estimates show that even Jupiter's force of gravity is less than one hundredth of a percent of that of the Sun.

However, even exerting such a small influence over a relatively long period of time affects the Earth's movement (as they say, drop by drop the sea is drained). The Earth's orbit is distorted because of its mutual gravitational attraction with other planets in the Solar System, which has been occurring for about 25,000 years. Exact calculations show that the eccentricity of the Earth's orbit changes nonperiodically and it typically takes approximately 100,000 years for these changes to occur. Figure 1.22 illustrates how the eccentricity of the Earth's orbit will change in the next million years. Eccentricity will decrease over the course of roughly another 25,000 years, which will cause the Earth's orbit to become almost circular. Thereafter, eccentricity will begin to increase and reach an amount four times greater than its current one.

Influence on the Earth by the planets in the Solar System not only causes precession with a constant angle, but also a slow change in the Earth's axial tilt, also known as the obliquity of the ecliptic (Fig. 1.23).

The principle of precession is based on the fact that the direction of the Earth's axis turns as it rotates around the North Pole.

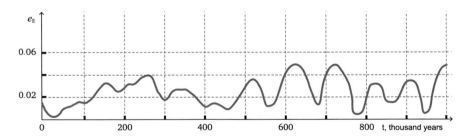

Fig. 1.22 Changes in the Earth's eccentricity in the future

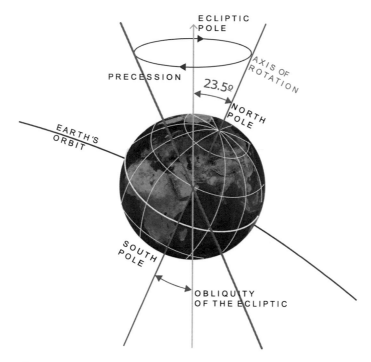

Fig. 1.23 The Earth's axis of precession

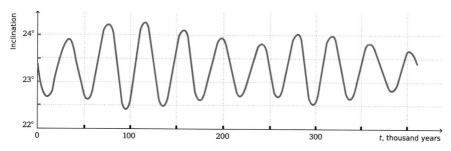

Fig. 1.24 Future changes in the Earth's obliquity

This angle of rotation is very small—about 20 s/years. It is clear that this vector will circumscribe the shape of a cone around the vertex after it has returned to its original position in about 26,000 years.

Figure 1.24 shows the calculations of this angle's measurement for the next 400,000 years.

The Earth's obliquity varies over an oscillation cycle of 41,000 years.

Influence of Planets on the Earth By influence we mean the ratio of the force of gravitational interaction of the Earth and a planet to the force of the gravitational interaction of the Earth and the Sun. This estimate yields:

$$\frac{F_{pl}}{F_S} \approx \frac{M_{pl}}{M_S}\left(\frac{R_{S-E}}{R_{pl-E}}\right)^2,$$ (1.33)

where $M_S = 11.99 \times 10^{30}$ kg, $R_{S-E} = 1.49 \times 10^8$ km and R_{pl-E} is the distance between the Sun and the Earth and between a planet and the Earth, respectively. Jupiter is the heaviest planet in the Solar System. Its mass is 3200 times more than the Earth's mass and is equal to $M_J = 1.91 \times 10^{27}$ kg. The distance from Jupiter to the Sun is 5.2 times greater than the distance from the Earth to the Sun and is equal to 7.75×10^8 km. By substituting numerical values, we see that $\frac{F_J}{F_S} \approx 5 \times 10^{-5}$, i.e., less than one hundredth of a percent.

Let's conduct a similar estimate for Venus. Although it is much closer to Earth, its mass is much less than that of Jupiter and about 20% less than the Earth's mass—$M_V = 4.83 \times 10^{24}$ kg. The distance from Venus to the Sun is equal to 1.07×10^8 km. In actuality Venus' influence is even weaker:

$$\frac{F_V}{F_S} \approx 3 \times 10^{-5}.$$ (1.34)

The mutual gravitational attraction of planets is strongest at the time that they are in conjunction (see Fig. 1.21) when they are in alignment on one side of the Sun. Since planets move periodically, they are in conjunction at regular intervals. Let's say there are two planets with orbital periods around the Sun T_1 and T_2. All planets revolve around the Sun in the same direction, and the relative frequency (i.e., the frequency ratio) is greater the closer the planet is to the Sun. The relative frequency of conjunction is the difference in the rate of rotation of two planets, i.e., $T_{conj}^{-1} = \left|T_2^{-1} - T_2^{-1}\right|$. Therefore, the conjunction period will be equal to:

$$T_{conj} = \frac{T_1 T_2}{|T_1 - T_2|}$$ (1.35)

After each planetary conjunction, a distortion of the planets' orbits occurs. This ratio distortion can be estimated as the ratio of the force of gravitational pull of a given planet to the nearest planet, $F_{pl1-pl2}$, and the force of gravitational pull of a given planet to the Sun, F_{pl1-S}. This means that the time during which a planet's orbit is significantly distorted due to mutual gravitational attraction with another planet is

$$T_{dist} \approx T_{conj}\frac{F_{pl1-pl2}}{F_{pl1-S}} = T_{conj.}\frac{M_S(R_2 - R_1)^2}{M_2 R_1^2}.$$ (1.36)

This estimate means that distortion of the Earth's orbit will take place after roughly 25,000 years and what is more, Jupiter and Venus have the greatest influence on the Earth.

How can the force of gravity affect a specific person on a specific day? Moreover, according to astrologers, gravity's influence affects different people in different ways: one day Venus helps a person deal with their work, another person gets help with their family matters and a third person receives financial assistance. When looking at a time series graph with a scale of 50,000 years, one can tell that these statements are obviously absurd. The universe functions according to completely different time and spatial categorial concepts than we do; it is undeniably much too massive to affect successfully obtaining a loan or passing a test in school. Truth be told, an icicle hanging from the roof of a tall building can have an incomparably greater impact upon a pedestrian than Mars in the "5th house."[1]

1.4   The Rotation of the Earth Around Its Axis

The rotation of the Earth around its axis can be illustrated by observing the diurnal motion of stars around the Celestial Pole, which is a fixed point on the celestial sphere. The North Pole is located within the constellation Ursa Minor, while the South Pole is in the constellation Octans. The picture (Fig. 1.25) taken in the Himalayas with a 24-h exposure shows the stars' visible rotation around the North Pole.

> The ecliptic passes through 13 constellations. Twelve of them are the zodiac constellations and the Ophiuchus constellation, all of which everyone knows well. You have heard of them. But it is important to remember that no reason or system has been used to divide the starry sky into constellations. The ancient Greeks adopted these divisions from the Babylonians. In ancient China, however, stars were united into constellations in an altogether different way.

It is necessary to devise an analogy with geographic poles, i.e., points at which the Earth's axis intersects with its surface.

Throughout the entire year, the Sun moves around the stars along an imaginary line called an ecliptic. This annual motion of the Sun is naturally only a reflection of the fact that the Earth revolves around it.

---

[1] The fifth house in astrology is the House of Pleasure. It describes the activities that bring joy, creativity and entertainment into one's life. It is ruled by the Sun and the sign of Leo. The position of the planets in the fifth house determine the effect they will have on a person's life (translator's note).

Fig. 1.25 Stars' rotation around the North Pole. This picture was taken in the Himalayas with a 6-h exposure

The Earth's obliquity of the ecliptic controls the seasonal cycle. Unfortunately, a large number of intelligent adults think that seasonal change is due to the Earth's rotation around its axis and that the Earth is closer to the Sun in summer and further away from it in winter.

Let's take a look at Fig. 1.26. The Earth's position on the extreme right aligns with December when the Northern Hemisphere is experiencing winter. Imagine that the Earth is rotating around its axis and we are standing at the North Pole. Note the fact that the Sun's rays do not fall at all on the area of the Earth's surface around the North Pole (this area is covered by the Arctic Circle). This means that right now we are at the North Pole during the polar night. The opposite is true near the South Pole where even at night the Sun does not set; at that time there is the polar day. As the Earth moves in orbit, the areas affected by the polar night in the north and the polar day in the south become fewer.

On March 20th, which is the day of the vernal equinox, the Earth's axis is perpendicular to the Sun's rays. On this day, the length of day and night is almost equal everywhere on the Earth's surface. Astronomical spring is coming. As the Earth continues to move in orbit, summer arrives in the Northern Hemisphere and the polar day begins. June 20th or 21st is the day of the summer solstice and the beginning of astronomical summer. At noon on these days, the Sun is at its zenith at a latitude known as the *Tropic of Cancer*.

The Earth continues to travel in orbit, and in the Northern Hemisphere fall arrives. On September 22nd or 23rd, the Earth's rotation axis is once

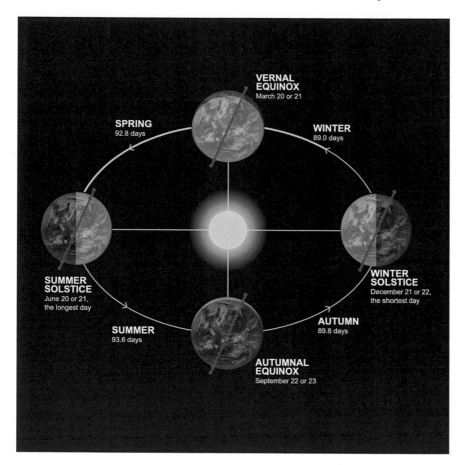

Fig. 1.26 The change of seasons

again perpendicular to the Sun's rays and day and night are exactly the same length. Finally, December 21st or 22nd is the day of the winter solstice. This is the beginning of astronomical winter. At noon on these days, the Sun is at its zenith at a latitude called the *Tropic of Capricorn.* Interestingly, the first and second half of the year is asymmetrical: the number of days from the spring to the fall equinox is 186, which is more than six months. But, on the other hand, less than six months pass from the fall to the spring equinox. The reason for this is that the Earth travels in an elliptical orbit and covers a greater distance in summer than in winter.

Due to the fact that the Earth does not rotate only around the Sun, but also around its axis, the time of sunset depends on geographical longitude. The Sun rises at the same time (without taking into account the surrounding terrain) in the places that are on the same longitude. This means that each one has its own local time, which is the time according to Greenwich Mean Time, plus the time required for the Earth to rotate around its axis at the necessary angle. This time is determined by using the following formula

$$t_{local} = t + \lambda \frac{1\,h}{15°}.$$ (1.37)

The local time at a given longitude λ is measured from the time of astronomical midnight at a given point. For the sake of convenience, local time is rounded up to an integer of hours in order to differentiate it from Greenwich Mean Time.

1.5 Inertial Forces

At the very beginning of this book, we talked about the geocentric model of the world, according to which the Earth is the motionless center and the whole world revolves around it. Maybe this is really true?

No, it's not. If the Earth did not rotate around an axis, we could see this without even looking at the sky. To the eye of an observer, a ball released from the center of a disk that rolls toward the disk's edge will roll in a straight line and at a steady pace, which is completely in line with Newton's laws. If we are on a rotating surface and are unaware of this, we won't be able to explain some of the oddities that have to do with a violation of Newton's laws. A case in point is that no force is acting on this ball, but yet it picks up speed somewhere off to the side. But moving bodies on the Earth's surface behave this same way, which once again proves that the Earth does indeed rotate. Both a stone thrown up and a shell fired along a meridian behave this same way. In order to understand what happens in these circumstances, it is easier to analyze physical phenomena within the context of a frame of reference that has to do with the rotating surface of the Earth. But frames of reference that are rotating with non-fluctuating angular velocity are non-inertial. This means that using Newton's laws in connection with these frames of reference requires some caution.

Now we will answer our previous question. From the point of view of two observers—one of whom is on the Earth and the other is off it—what will the path of motion be of a projectile fired from a cannon on the North Pole? If the Earth did not rotate around its axis, the projectile's flightpath would be in the meridian plane, that is, the projected path onto the Earth's surface would align with the meridian line.

But the Earth rotates and the fixed meridian plane revolves with the Earth from west to east. From the point of view of a motionless observer located off the Earth, the plane in which the shell flies remains motionless.

But what does the observer see who is at a specific point on the Earth's surface? They see a fixed meridian line and a flying shell that deviates from it to the west. For both the first and second observers, the shell will deviate from the meridian line that is rotating with the Earth; in other words, both observers will see the same thing, although they view this motion using different frames of reference.

It is thought that in such situations, the lateral deviation of the shell is caused by some force that acts only in the non-inertial frame of reference, although we understand that the "real" reason that the shell does this is because the Earth rotates. These extraordinary forces are called *inertial forces*. They, unlike ordinary forces, are not the result of the interaction of any two objects.

In a rotating coordinate system, two such inertial forces act on any object. The first force is centrifugal. Its value depends on the distance from a given point to the axis of rotation. The second force is the Coriolis force. Its magnitude depends on an object's speed with regard to the rotating frame.

Let's first analyze the centrifugal force.

Centrifugal Force To begin, let's answer this question: Why does a boy's cap fly off his head when he is riding on a carousel (Fig. 1.27), if the rotational velocity exceeds a certain value? From the point of view of an observer who is riding on the carousel with the boy, the explanation is as follows: the boy does not lose his cap due to the friction force F_{frict}. In addition, the inertial force acts on it, which is equal to:

$$F_I = m\omega^2 r, \tag{1.38}$$

where m is the mass of the cap, ω is angular velocity and r is the distance to the rotation axis.

If the condition $F_I < F_{\text{frict}}$ is fulfilled, then the boy's cap does not fly off his head. If the rate of rotation increases and at a certain point the inertial force exceeds the friction force $F_I > F_{\text{frict}}$, his cap will fly off. From the point of view of a motionless observer standing on the Earth next to the carousel, only one force acts on the cap and it is the friction force. This force provides the cap with centripetal acceleration. Since the force of static friction increases to a certain limit, with an increase in the rate of rotation, the cap will come off the boy's head at a certain moment and begin to move in a steady motion. Then it will fly off the carousel.

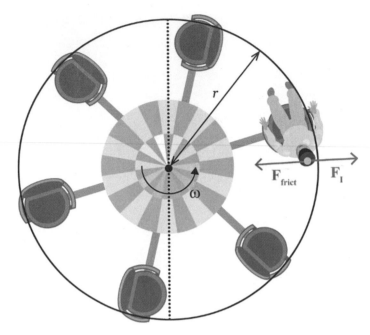

Fig. 1.27 Forces acting upon the cap of the boy who is riding on a carousel

Coriolis force Now let's move on to an analyzation of the Coriolis force. Imagine a person is riding on a carousel and moving from seat to seat (Fig. 1.28).

The angular velocity of the carousel is ω, its radius is r and the velocity of the person in the frame of reference corresponding to the carousel is equal to v_0 and is channeled in the direction of rotation. In a fixed reference, the person's total velocity is $v = \omega r + v_0$. The inertial force that acts on the person is equal to the product of their mass m by the centripetal acceleration $\frac{(\omega r + v_0)^2}{r}$. This force is balanced by the reaction force R, which the person's seat exerts on them.

In a rotating frame related to the carousel, the force that acts on the person is $\frac{mv_0^2}{r}$. Then Newton's second law looks like this:

$$\frac{mv_0^2}{r} = R + F_{\mathrm{I}}, \tag{1.39}$$

where $F_{\mathrm{I}} = -m\omega^2 r - 2m\omega v_0$.

We see that the inertial force has two components. The first one is the centrifugal force, which we know well. The faster the carousel's rotational velocity and the greater the distance to the center of rotation, the stronger

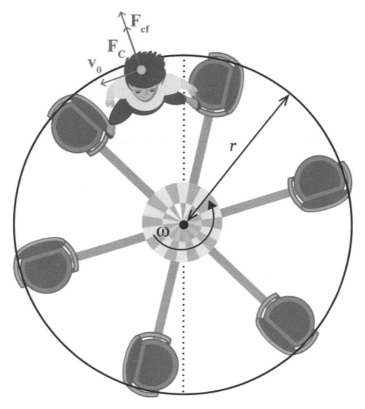

Fig. 1.28 Forces acting on a person who moves from seat to seat on a carousel

this force is. The second component is the Coriolis force. It does not depend on a person's location, but is proportional to their velocity. The direction of the Coriolis force is perpendicular to the rotation axis and an object's velocity.

This means that if a person moves around on a carousel changing seats, that is, their velocity is directed along the tangent toward the circumference, then the Coriolis force is directed along the radius of the circumference. If a person moves from the center of the carousel toward the periphery, then the Coriolis force will push them to the side because it is being channeled perpendicular to the radius.

The direction of the Coriolis force is determined using the left-hand rule, as shown in Fig. 1.29.

But what will the Coriolis force do if it is acting on an object in a rotating frame of reference and the object's velocity is channeled in a random way? If an object's velocity forms an arbitrary angle with the rotation axis, then it is necessary to consider only the projection of velocity onto a plane that is perpendicular to the axis of rotation (see Fig. 1.29). The Coriolis force will

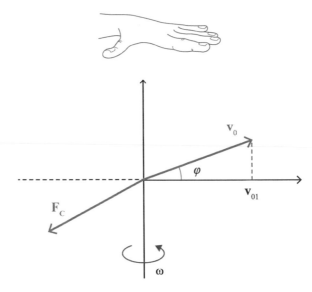

Fig. 1.29 The left-hand rule used to determine the direction of the Coriolis force

then be $-2m\omega v_0 \cos\varphi$. If we introduce the vector $\boldsymbol{\omega}$ (its value is equal to the angular velocity and we have chosen the direction of the rotation axis so that rotation occurs in a clockwise direction), then the Coriolis force will be perpendicular to the plane on which the vectors $\boldsymbol{\omega}$ and $\mathbf{v}_0$ lie. In other words, the Coriolis force can generally be written in the form of a vector product:

$$\mathbf{F}_C = 2m[\mathbf{v}_0 \times \boldsymbol{\omega}]. \tag{1.40}$$

One sees the Coriolis force at work on the surface of the Earth when observing Foucault's pendulum, which is a heavy ball suspended on a fairly long thread. If the pendulum becomes unbalanced, it starts to oscillate. In this case, the plane of vibration does not remain constant, but slowly rotates around the line in a clockwise direction (if you look at it from above). When in Russia, one can see a Foucault pendulum with the longest suspension at the Moscow Planetarium.

In order to explain the rotation of the pendulum's plane of vibration, let's suppose that the Earth rotates at a steady pace with angular velocity. For simplicity's sake, the pendulum is suspended at the North Pole (Fig. 1.30).

From the point of view of a motionless observer looking at the Earth from outer space, the pendulum's plane of vibration is motionless and the Earth rotates counterclockwise. An observer with a frame of reference that is rotating together with the Earth will assert that the rotation of the pendulum's plane of vibration is due to the Coriolis force. If the pendulum does not

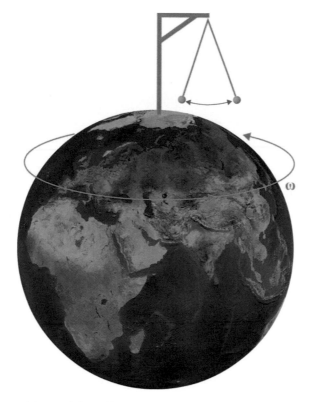

Fig. 1.30 Foucault's pendulum illustrating the Coriolis force

significantly deviate from a state of equilibrium, then the speed of the ball is at any time perpendicular to the Earth's rate of angular motion ω_E. Consequently, the Coriolis force always lies on the horizontal plane and is normally channeled toward the velocity of the ball; if we look toward the direction of the ball's movement, then this is to the right. The Coriolis force depends on geographical latitude. If the pendulum is suspended at a point on the Earth's surface with the geographical latitude θ, then the angular velocity will be equal to $\omega = \omega_E \sin\theta$, and the Coriolis force will be proportional to $\sin\theta$. Since the swing speed of the pendulum's plane of vibration disappears at the Equator, the Coriolis force is equal to zero there and increases in absolute value as one gets closer to the poles.

By way of illustration, we will consider some important aspects of the Coriolis force.

Deviation of Free-Falling Objects From the point of view of an observer bound to an inertial frame of reference, an object thrown down from the height H will fall at the horizontal velocity $v = \omega_E(H + R_E)\cos\theta$. When an object is falling, it retains this velocity; in its horizontal motion, it overtakes stationary points on the Earth's surface. Since the Earth rotates from the west to the east and the object is overtaking the Earth, this object will deviate from the vertical line to the east. The displacement of this object falling eastward will be equal to

$$\Delta L = \frac{1}{3} \frac{\omega_E \cos \theta v_{end}^3}{g^2}, \tag{1.41}$$

where $v_{end} = \sqrt{2gH}$ is the object's vertical velocity at the moment that it comes in contact with the Earth. In the Southern Hemisphere, falling objects deviate westward.

A projectile that was fired from a cannon is also a falling object. Let's say that you are an artillerist and you need to evaluate the correction for the Coriolis force if a shell is fired along a parallel at a speed of 1 km (0.62 mi)/h. What will its deviation be? If we do not take air resistance into account, the shell's deviation will be several dozen meters (several hundred feet)!

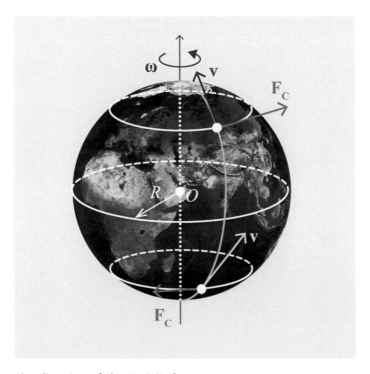

Fig. 1.31 The direction of the Coriolis force

Effect of the Coriolis force on the Flightpath of an Artillery Shell When a shot is fired from a gun that is pointing north, it will deflect eastward in the Northern Hemisphere and westward in the Southern Hemisphere (Fig. 1.31). When a shot is fired along the Equator to the west, the Coriolis force presses the shell against the Earth and increases the rate at which it falls. However, if the shot is fired to the east, the Coriolis force slightly lifts up its flightpath and decreases the velocity at which it falls. In this situation, when $m_{shell} = 1\,kg\,(2.2\,lbs)$, the Coriolis force is equal to $F_C = 0.15\,N$. If the flightpath of the shell is, for example, $\Delta t = 30\,s$, one can estimate the lateral momentum change $\Delta p_y \approx F_C \Delta t = 4.5\,N\,s$. Then the incremental velocity in the lateral direction is $\Delta v_y \approx 4.5\,m\,(14.76\,ft)/s$ and the lateral deviation is $\Delta L \approx \Delta v_y \Delta t = 65\,m\,(213.26\,ft)$. When a shot is fired along a meridian with the longitude λ, the incremental velocity can be estimated as $\Delta v_y \approx 4.5 \sin \lambda$ m/s, while the lateral deviation is $\Delta L \approx 65 \sin \lambda$ m. But if, for example, we are dealing with a rocket that is moving at a velocity of 500 m (1640.42 ft)/s for 3 min, then because of the Coriolis force the lateral deviation can reach 1 km (3280.84 ft). This difference must be taken into consideration.

Non-symmetrical Wearing Down of Left and Right Railroad Tracks (of trains traveling in one direction). The Coriolis force is small. For example, it is less than a tenth of a percent of the force of gravity for a car traveling at a velocity of 60 km (37 mi)/h. But for protracted periods of movement the effect of the Coriolis force can be significant. For example, in the Northern Hemisphere railroad tracks for trains that run to the right wear out faster than tracks for trains that run to the left, while in the Southern Hemisphere the opposite is true.

Coriolis Force Acting on a Train Let's assume a train having a mass of $M = 150\,T$ travels along a meridian northward with a velocity of $v = 72\,km\,(44.74\,mi)/h$ at the latitude of Moscow $\varphi = 56°$. The angle between the angular velocity vector of the Earth's daily rotation and the tangent to a meridian is equal to the site latitude (Fig. 1.32).
The Coriolis force will be equal to:

$$F_C = 2Mv\omega_E \sin \varphi = 362\,N. \qquad (1.42)$$

This force corresponds to the weight of cargo having a mass of $\Delta M = \frac{F_C}{g} = 37\,kg\,(81.57\,lbs)$, which is 2.5×10^{-4} of the train's mass.

Non-symmetrical Erosion of River Banks. Since the Coriolis force presses water to the right bank in the Northern Hemisphere, the friction force between the water and the soil is stronger there. Because of this, water circulation and the transfer of soil particles from the right to the left bank occurs.

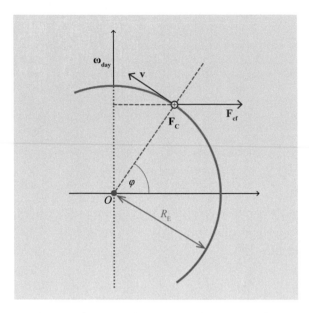

Fig. 1.32 Forces acting on the train that moves along a meridian

Therefore, the right side of riverbanks is steeper and more eroded than the left one (Fig. 1.33).

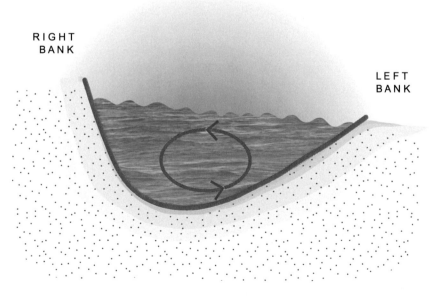

Fig. 1.33 Difference in the steepness of left and right riverbanks in the Northern Hemisphere

Swirling Winds over the Earth's Surface and Ocean Currents The Coriolis force strongly affects the movement of major air streams and water flow, i.e., wind and ocean currents. We will discuss what causes trade winds to form a bit later. First and foremost, it is important to note that if the Earth did not rotate, the air masses that get warmed up in the equatorial zone due to convection would move northward in the Northern Hemisphere and southward in the Southern Hemisphere. While moving away from the Equator at a high altitude, the air masses cool off and settle down.

While the wind is forming wind cells (Fig. 1.34), it then turns back to the Equator at a low altitude. Because of the Coriolis force, in the Northern and Southern Hemispheres these "low" winds turn westward. These are trade winds. Let's assume that an air mass travels the distance $L = 1000\,\text{km}\,(621.37\,\text{m})$ toward the Equator at an average velocity of $v = 10\,\text{m}\,(621.37\,\text{ft})/\text{s}$. Then during the travel time $t = \frac{L}{v}$ the wind will be westerly $v_w \approx 2\omega L \sin\lambda \approx 10\,\text{m}\,(32.81\,\text{ft})/\text{s}$. The Coriolis force radically changes the global picture of atmospheric winds.

This force causes underwater ocean current to rotate. We will elaborate on this interesting phenomenon in the respective section about water.

Orbit of the Earth's Satellites A satellite rotates around the Earth due to gravity in a closed trajectory known as an ellipse. The plane of this ellipse is fixed with relation to the stars. However, the Earth also rotates around its axis. Let's assume the satellite's period of revolution around the Earth is T_{sat}. During the time T_{sat} while the satellite is making one revolution, the Earth will rotate at the angle

$$\Delta\lambda = 2\pi\frac{T_{\text{sat}}}{T_{\text{E}}}. \tag{1.43}$$

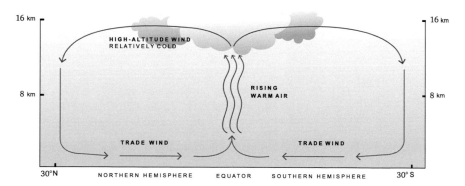

Fig. 1.34 Wind cell formation

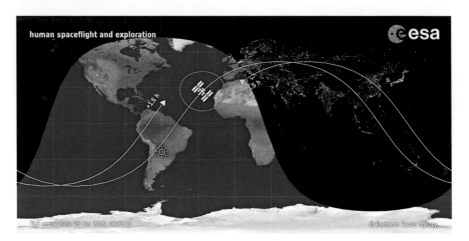

Fig. 1.35 The path of motion of a satellite on the screen at Mission Control Center

It is precisely for this reason that in a rotating coordinate system connected with the Earth's surface, a satellite's orbit resembles an open curve as can be seen on the screens at Mission Control Center (Fig. 1.35).

If a satellite is in the Northern Hemisphere and is moving eastward, the Coriolis force deflects it to the south. In the Southern Hemisphere, the Coriolis force changes signs and begins to make satellites return to the north. In a fixed coordinate system corresponding to the stars, satellites obviously move in a closed trajectory. It is only our desire to "connect" them to a rotating frame that forces us to use the Coriolis force to correctly describe motion in a coordinate system pertaining to the Earth. In the Northern Hemisphere, this force acts to the right in the direction of motion, while in the Southern Hemisphere it acts to the left.

1.6   The Calendar

We all use calendars in our everyday life, but when we plan our lives and fill in our daily planners, we often do not give any thought to where calendars came from and why they function as they do. People came up with calendars. Dividing the year into 12 months was simply a convention, just as was deciding that a week has seven days. It would have also been possible for a year to last three months with each month consisting of two weeks. It is not hard to calculate the length of a week; without question, even more over-the-top alternatives than seven days are possible. But whatever type of calendar there is, no one can deny that a year is equal to the Earth's revolution period around the Sun. As we know, this is determined by Kepler's third law, according to which the Earth rotates around its axis with an angular velocity of $7.29 \times 10^{-5}\,\text{s}^{-1}$ and its axis is tilted toward the plane of rotation at an angle of $23°\ 27'$.

The time-reference system associated with astronomical phenomena dates back to ancient Babylon. After Greek troops led by Alexander the Great conquered Babylon in 331 BC, they adopted this system. In accordance with it, whole days were divided into 24 h with 2 h for each zodiac constellation. Since the Sun moves about $1°$ per day along the ecliptic, in ancient times people divided a circle into $360°$, an hour into 60 min and a minute into 60 s.

> The precession of the Earth's axis is another important astronomical discovery that was made by the Greek astronomer Hipparchus. He noticed that the equinox occurs a little earlier than when the Earth makes a full revolution in orbit with respect to the stars. This is because the Earth's axis rotates slowly, revolving around pole c with a period of about 26,000 years. The angular velocity of the precession of the Earth's axis is $\omega_{\text{prec}} = 7.72 \times 10^{-12}$ s. In accordance with this, the precession period $T_{\text{prec}} = \frac{2\pi}{\omega_{\text{prec}}} = 2.578 \times 10^4$ years $= 8.135 \times 10^{11}$ s.

The creators of the Gregorian calendar, which is the calendar used today, devised it so that the day of the spring equinox would always fall on the same date—March 21st. For this reason, our calendar is based on the change of seasons and of day and night, which is connected to the Earth's revolution around the Sun and around its axis. This is called a *solar calendar*. The physical quantities that are needed to construct such a calendar are the following: the Earth's revolution period in its orbit around the Sun— $T_E = 3.155815014 \times 10^7$ s and the Earth's revolution period around its axis— $P_E = 8.616409 \times 10^4$ s.

Now let's think how we can determine the length of a year. It would seem that this is very simple—we need to find the number of 24-h periods between, let's say, the days of the spring equinox. But it is not that simple because there are a few problems. First, precession is interfering and, second, the length of a year is not equal to the total number of 24-h days.

Let's determine the time interval, known as the tropical year T_{trop}, between two consecutive moments of the spring equinox. It is specifically the tropical year upon which the calendar is based because then each year the spring equinox falls on the same date as the four seasons. Thus, the moment of equinox corresponds to the point when the direction from the Earth to the Sun is perpendicular to the Earth's axis. Because of precession, this point moves along the ecliptic in a direction that is opposite to the movement of the Earth with the period T_{prec}. Since planets' rotation and the movement of the equinoctial point are opposite, the corresponding frequency ratios must be added

$$T_{trop}^{-1} = T_E^{-1} - T_{prec}^{-1} \tag{1.44}$$

and the period must be determined

$$T_{trop} = \frac{T_E T_{prec}}{T_E + T_{prec}} = 31{,}556{,}926\,\text{s} = 365\,\text{days}\,5\,\text{h}\,48\,\text{min}\,46\,\text{s}. \tag{1.45}$$

Because of precession, a tropical year is 1224 s shorter than the actual period of the Earth's revolution around the Sun. Now let's calculate the average number of sunny days per year. Why is it correct to talk about average days? The reason is that the Sun moves sporadically along the ecliptic throughout the year due to the elliptical shape of its orbit. Since the Earth revolves around the Sun and rotates around its axis in one direction, the average frequency ratio P_0^{-1} (this is a day) of the Sun reaching its zenith is neither more nor less than the difference between P_E^{-1} and T_{trop}^{-1}:

$$P_0^{-1} = P_E^{-1} - P_{trop}^{-1}, \; P_0 = \frac{P_E \times T_{trop}}{T_{trop} - P_E} = 86{,}400\,\text{s} \tag{1.46}$$

which is equal to the number of seconds in a day: $24 \times 60 \times 60$.

We can calculate how many days there are in a tropical year:

$$\frac{T_{trop}}{P_0} = 365.2422. \tag{1.47}$$

You have probably learned why once every four years, a leap year occurs and there are 366 days instead of 365. Julius Caesar first introduced a leap year in ancient Rome with a frequency of four years, which is why the calendar came to be called the *Julian calendar*. The average length of the calendar year of the Julian calendar is, consequently, 365 days and 6 h, which is more than the actual length of the tropical year by 11 min and 15 s.

However, adding leap years does not guarantee that the spring equinox will always occur on the same date because after $(365.25 - 365.2422) - 1 = 128$ years the date of the spring equinox would shift by one day. This is due to rounding 365.2422 up to 365.25. Hence, by the end of the fourteenth century the day of the spring equinox had fallen back an entire 12 days on the calendar and for that reason, upon the recommendation of Pope Gregory XIII, a process to reform the calendar was initiated in 1582. The new calendar, which was called the Gregorian calendar, was different from the Julian calendar because of an additional rule that called for alternating regular and leap years. Years that can be divided by four were designated as leap years; the exception being years that end with two zeros and that cannot be divided by 400.

Owing to this, in the Gregorian calendar, which we use, for every 400 years there are not 100 leap years, but instead 97. The year 1600 was a leap year, whereas 1700, 1800 and 1900 were not; 2000 was again a leap year. Moreover, the calendar was shifted by 10 days, which is why the day after October 4th, 1582 suddenly became October 15th. Since then, differences in the Gregorian and Julian calendars have increased by another three days: by one day in 1700, one in 1800 and one in 1900. Until 1918 the Julian calendar was used in Russia, but on January 26th, 1918 the Gregorian calendar was introduced.

Thus, the average length of the year according to our calendar is equal to:

$$\left(365\frac{97}{400}\right) \text{days} = 365 \text{ days } 5\,\text{h } 49\,\text{min } 12\,\text{s}. \tag{1.48}$$

This is more than the actual length of the year by 26 s. We see that even such an improvement as this does not fully account for the precession of the Earth's axis; however, since the corrections are small, our calendar can most certainly be used for at least several thousand years.

It is possible to come up with other calendars that would make for the length of the calendar year much closer to that of the tropical year than the Gregorian calendar does. Continued fractions, which are a type of mathematical construction, make this calculation possible. We have already seen that the average length of the Julian yr is 365 1/4 days, while the average length of the Gregorian calendar is 365 97/400 days. As a result, both the Julian and Gregorian calendars have a cycle of 4 years and 400 years, respectively. A new calendar with the cycle length q can serve as an adequate approximation for the length of a tropical year, which takes the form $365\frac{p}{q}$. Here p and q are positive integers; moreover $p < q$. It has been proven in mathematics that any positive number A can be uniquely decomposed into a fraction:

$$A = a_0 + \cfrac{1}{a_1 + \cfrac{1}{a_2 + \cfrac{1}{a_3 + \ddots}}}, \tag{1.49}$$

where a_0 is the integral part of A, but a_1, a_2, a_3 are all positive integers.

We will write the number 10/7 in the form of a continued fraction:

$$\frac{10}{7} = 1 + \frac{3}{7} = 1 + \cfrac{1}{\frac{7}{3}} = 1 + \cfrac{1}{2 + \frac{1}{3}} \tag{1.50}$$

Hence, it is possible to find a common "recipe" in order to create a series of calendars with a year-long length that is much closer to the length of a tropical year. We can decompose the length of time that a tropical year has sunny days into a continued fraction in the following way:

$$365\frac{52{,}313}{216{,}000} = 365 + \cfrac{1}{4 + \cfrac{1}{7 + \cfrac{1}{31 + \cfrac{1}{236 + \cfrac{1}{926 + \cfrac{1}{79 + \cfrac{1}{7 + \ddots}}}}}}} \tag{1.51}$$

Each of the numbers: 365, $365 + \frac{1}{4} = 365\frac{1}{4}$; $365 + \cfrac{1}{4 + \frac{1}{7}} = 365\frac{7}{29}$, $365 + \cfrac{1}{4 + \cfrac{1}{7 + \frac{1}{1}}} = 365\frac{8}{33}$, $365 + \cfrac{1}{4 + \cfrac{1}{7 + \cfrac{1}{1 + \frac{1}{3}}}} = 365\frac{31}{128}$, etc. represents the basis for its "own" calendar.

In Table 1.2, the first line describes the Julian calendar, which has four years in a cycle and one year is leap year. No one suggested using a calendar based on the second line in the table. However, Omar Khayyam, who was not only a poet, but also an astronomer and a mathematician, proposed adopting a calendar based on the third line with eight leap years every 33 years. Finally, in the nineteenth century the German astronomer von Mädler suggested using a calendar based on the fourth line. That calendar is much more precise than our Gregorian calendar and also much simpler—its cycle consists of 128 years and not 400. It is surprising that von Mädler's calendar is much more accurate than the Gregorian calendar—it gives an insignificant amount of error of 1 s per year. Using the properties of continued fractions, one can prove that the calendar Khayyam proposed adopting is the most accurate of

Table 1.2 Possible calendars based on numbers produced by continued fractions

Number	Length of a calendar year	Discrepancy
365 1/4	365 days 6 h 0 min 0 s	− 11 min 15 s
365 7/29	365 days 5 h 47 min 35 s	1 min 10 s
365 8/33	365 days 5 h 49 min 5 s	− 20 s
365 31/128	365 days 5 h 48 min 45 s	< 1 s

all calendars with a cycle of less than 33 years, while the calendar that von Mädler recommended is the most accurate of all calendars with a cycle of less than 128 years. In Mädler's calendar an error in one day adds up over 100,000 years! Why isn't the Gregorian calendar in our table? The reason is simple: in the fourteenth century astronomers used inaccurate information having to do with the length of a tropical year. Although there was only a 30 s deviation from the true value, this turned out to be enough to adopt the less-than-perfect calendar that we use to this day.

Further Reading

1 Bickerman, E.J.: Chronology of the Ancient World. Cornell University Press, New York (1980)
2 Byalko, A.V.: Our Planet the Earth. MIR Publisher, Moscow (1983)
3 Feynman, R.: The Character of Physical Law. MIT Press, Cambridge (2017)
4 Judge, P.: The Sun: A Very Short Introduction. Oxford University Press, Oxford (2020)
5 Moritz, H.: Earth Rotation: Theory and Observation. UNKNO (1987)
6 Murray, C.D., Dermott, S.F.: Solar System Dynamics. Cambridge University Press, Cambridge (1999)
7 Pannekoek, A.: A History of Astronomy. Dover Publications, New York (2011)
8 Richards, E.G.: Mapping Time: The Calendar and Its History. Oxford University Press, Oxford (2000)
9 Varlamov, A.A., Aslamazov, L.G.: The Wonders of Physics, 4th edn. World Scientific, Singapore (2019)

2

The Earth and the Moon

Abstract The physical phenomena conditioned by the Earth–Moon interaction are discussed in this chapter. We consider the color of the Moon as it is seen from the Earth, the rules of Solar and Lunar eclipses, the origin of tidal effect due to the Moon's influence on the Earth, and the physics of high and low tides on the Earth. Additionally, we study the precession of the Earth's axis and the deceleration of the Earth's rotation caused by the Moon.

For as long as anyone can remember, the Moon, clearly standing out in the starry sky, has attracted people's attention. Our ancestors tried to see it more clearly through the first telescopes and by sending the first spacecraft to the Moon. It is the only celestial body on which a human being has set foot. On September 13, 1959 the first spacecraft—the Soviet Union's "Luna 2" (Fig. 2.1)—reached the Moon. Ten years later in 1969, the American spacecraft "Apollo 11" landed on the Moon and Neal Armstrong was the first person in the history of mankind to step foot on it (Fig. 2.2). So just what is the Moon? Where did it come from? What kind of influence does it exert on the Earth?

The Moon is so large and so close to us that the amount of sunlight it reflects during the full phase provides a considerable amount of light at night when there isn't any cloud cover. The Moon has a large enough mass to influence the shape of the Earth. This influence is most visible in the Moon's control of high and low tide in seas and oceans. The Moon is the key factor

© The Author(s), under exclusive license to Springer Nature
Switzerland AG 2023
D. Livanov, *The Physics of Planet Earth and Its Natural Wonders*,
https://doi.org/10.1007/978-3-031-33426-9_2

Fig. 2.1 The Soviet Lunar Orbital Station "Luna 2"

Fig. 2.2 The first person on the Moon

that causes a change in the position of the Earth's axis of rotation, i.e., its precession. Moreover, a change in the Earth's shape because of the Moon's influence on it is followed by a specific deceleration of its rotation, which slowly lengthens the duration of each day. Finally, in certain places and at certain times the Moon covers the Sun, which results in a solar eclipse. This is how our nearest neighbor in space lets us know that it is there. We will now explore all of these phenomena.

Out of all the planetary satellites of the Solar System, the Moon has the greatest influence on the planet it orbits—the Earth. The Moon is always facing Earth on one side, which is because of deceleration caused by tidal forces. The Moon rotates around its axis with the same frequency and in the same direction in which it rotates around the Earth. It is curious that the center of the Moon's mass is heavily shifted toward the Earth in relation to its geometric center. This means that in the "Earth–Moon" system tidal forces used to be much stronger than they are now. In other words, our planet and its satellite were closer to each other than they are today, which once again confirms the hypothesis that at some point the Moon and the Earth were one celestial body.

2.1   The Color of the Moon

The first thing that makes the Moon noticeable is, of course, its color! Without the Moon, nights would be dark and the night sky would not have its crowning glory.

First, let's answer this question: Why do we rarely see the whole Moon in the nighttime sky? It is obviously not because the Earth's shadow falls on it. Instead, it is due to something completely different: the Sun illuminates only one side of the Moon, but this illuminated side is not entirely visible from the Earth. It is only when the Moon is full that we see the side of it that the Sun illuminates. Figure 2.3 shows the Moon's phases: 1—new moon (the Moon is not visible); 5—full moon.

The giant-impact hypothesis is the most credible theory today that explains the Moon's formation. According to it, more than four billion years ago the Earth collided with some kind of gigantic celestial body. An explosion caused a piece of the Earth to break off and, subsequently, it became the Moon— the Earth's satellite. The mass of this "piece" turned out to be about 81.3 times smaller than the mass of the rest of the Earth. The fact that the Moon's

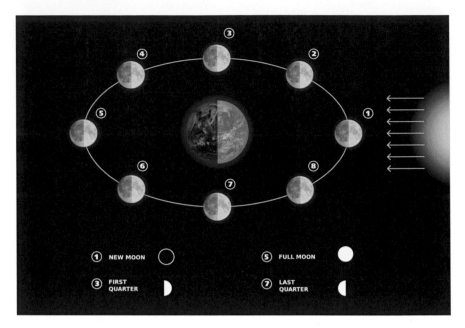

Fig. 2.3 The phases of the Moon

chemical makeup is similar to that found on the Earth is one of the factors that confirms this hypothesis.

In order to describe the Moon's glow, it is necessary to apply certain characteristics, which allow us to objectively define and compare the brightness of different celestial bodies. Indeed, some of them appear to be very bright, while others are barely visible (Fig. 2.4). The majority, however, are not visible at all to the naked eye.

In order to characterize the visibility of stars, the concept of stellar magnitude (or stellar brilliance) has been introduced. The ancient Greek astronomer Hipparchus divided all of the stars into six magnitudes. He assigned the brightest stars to the category of first-magnitude stars, while the dimmest were put in the category of sixth-magnitude stars. Hipparchus divided the rest in the range from one to six. As it was discovered later, the variation in how we visually perceive stars' brightness per unit actually means that there is a substantial difference in the physical characteristics of the flow of energy. This means that the lower the star's magnitude on the numerical scale, the brighter it is and vice versa. One cannot call this type of historically established system successful because the numerical scale turned out to be a logarithm that was turned upside down. For an ordinary person who is not

Fig. 2.4 The night sky

an expert in astronomy or mathematics, a logarithm like this is difficult to figure out. Therefore, there is no reason to be surprised when we learn that the Sun, which has a magnitude of $m = -26.7$ is 400,000 brighter than a full moon, which has a magnitude of $m = -12.74$.

The next brightest objects after the Moon are Venus (Fig. 2.5) and the International Space Station. One could say that mankind has brushed a small "stroke" on the majestic picture of the starry sky.

Both the Earth and the Moon shine light that has been reflected off the Sun. In order to characterize these light sources, we have the concept of *albedo*.

Fig. 2.5 The Moon and Venus in the night sky

Albedo is a percentage of solar radiation that is reflected by a surface.

The Earth's albedo is determined by the properties of its atmosphere and surface because one portion of the Sun's rays is reflected off clouds, while another portion bounces off the Earth's surface. It is clear that the albedo of the Earth's surface varies a great deal: the ice of Antarctica or the Arctic, so perfectly clear that looking at it hurts one's eyes, reflects up to 86% of light, while the sandy or rocky surface of the desert reflects less than 20% of light. It also makes sense that albedo rises in winter when snow falls and falls when it melts. Leafy trees also affect the albedo of the Earth's surface. In an attempt to solve global warming, some scientists have suggested that people paint pavement and house rooftops white. This solution has substance—white surfaces will increase albedo especially in metropolitan cities. Additionally, the average temperature in cities, which is several degrees higher than in the suburbs, will drop significantly.

Albedo varies depending on the change in the solar incidence angle: the greater the angle of incidence measured from the surface normal, the more energy from incoming radiation is reflected and the less that is absorbed. Conversely, albedo decreases when the Sun's height above the horizon drops; when the Sun is at its zenith, albedo reaches its minimum. Therefore, albedo changes depending on the time of day. The albedo of a water surface is on average lower than that of most natural land surfaces and depends on the solar incidence angle, the height of the Sun, the ratio of beam radiation and diffuse radiation in the sky, as well as swells in seas and oceans. When the Sun is at its zenith, the beam radiation that reflects off smooth water produces an albedo of 2%.

If you have flown in an airplane and sat in a window seat, it is possible that you may have seen how light reflects off different surfaces such as water, forests, farmland, etc. Some of these surfaces seem dark and have a lower albedo, while others seem light and have a higher albedo. It is interesting to observe water surfaces because when they are far off in the distance, they look like mirrors when we see them from the direction of the Sun because they reflect so much falling light. However, the surfaces that are directly below us seem like absolutely black spots. When there are swells in the ocean and foam and whitecaps form, the albedo of water increases. Diffuse radiation in the sky that reflects off the ocean's surface produces an albedo that ranges from 5 to 11%.

Thus, the average albedo of the Earth in the visible light spectrum is about 40%, while the total albedo, which takes into account the entire spectrum, is about 30%. When the sky is cloudless, the visual albedo of the Earth is about 15%, which means we can think of it as being approximately $A_E = 0.30$.

Things are much simpler as far as the Moon's albedo is concerned because the Moon lacks an atmosphere and, furthermore, its surface is dominated by igneous rocks that absorb the bulk of solar radiation. Hence, the Moon's albedo is only slightly different from the average albedo on the Earth, which is roughly equal to 8%, $A_M = 0.08$.

The average distance today between the Earth and its satellite is 3.84×10^5 km, which is 60 greater than the Earth's radius. The Moon has a radius of 1738 km (1079.94 mi) at the Equator. It would be safe for us to assume that the Moon moves around the Earth in an elliptical orbit with an eccentricity of 0.055. The Moon's orbit is close to the ecliptic plane, but it does not exactly coincide with it: the angle between the planes is about 5°.

Within the period of two to three nights just before a new moon, when the crescent moon is very narrow and still bright enough so that we can distinguish between its faint light and the rest of the Moon's disk, we can see an interesting phenomenon: there is a faint glow from the dark disk of the Moon that is quite noticeable against the black sky. This curious occurrence is called *earthshine* (Fig. 2.6).

Leonardo da Vinci made a sketch of earthshine (Fig. 2.7).

The path of light rays that give the Moon its ashen color is shown in Fig. 2.8. As you can see, a ray of light is reflected from the Earth's surface, then reflects off the Moon's surface and after a double reflection, this ray of light is then visible to an observer on the Earth. This glow is especially impressive during the new moon phase when the Moon is dark and all of the hemisphere—as it appears from the Moon—is illuminated by the Sun. It is interesting that although the brightness of earthshine is 24,000 times less than the brightness of the Moon when it is full, we are still able to clearly see it.

Brightness of Earthshine Let's estimate how much weaker the Moon's earthshine is than its ordinary light. The Earth evenly emits reflected sunlight in all directions with solar power P_E.

Fig. 2.6   Earthshine

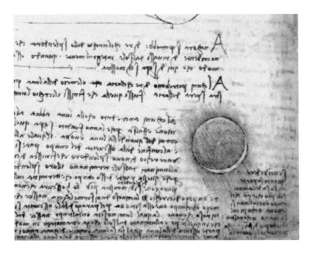

Fig. 2.7   A crescent moon with earthshine. Leo da Vinci's sketch from his book *Codex Leicester*

We will designate the portion of this stream of light that falls on the Moon as S_M. We have:

$$S_M = \frac{P_E}{2\pi R_{E-M}^2} \qquad (2.1)$$

Fig. 2.8 The path of light rays when one sees earthshine

Then the earthshine intensity is

$$P_{M(\text{earthshine})} = A_M S_M \pi R_M^2 = \frac{\pi}{2} \frac{A_E A_M R_E^2 R_M^2}{R_{E-M}^2} S. \qquad (2.2)$$

It is easy to calculate how much this intensity is less than the Moon's intensity when it is full:

$$\frac{P_{M(\text{earthshine})}}{P_M} = \frac{1}{2} A_E \frac{R_3^2}{R_{E-M}^2} = \frac{1}{24,000}. \qquad (2.3)$$

In this situation, the ratio for intensity is also true for brightness.

It is important to note that the Earth illuminates the Moon's surface in the center of its visible area approximately 450 times more than the Moon illuminates the Earth's surface! One can only imagine the breath-taking image of the Earth in the black moonlit sky—a huge bluish-white disk shining hundreds of times brighter than a full moon as we see it from the Earth. Seeing the Earth this way will most certainly be the main attraction for all of tomorrow's space travelers.

2.2   Solar and Lunar Eclipses

Since the Moon rotates around the Earth and the Earth then rotates around the Sun, it is possible that the Earth can be in the Moon's shadow or the Moon can be in the Earth's shadow. The former is called a *solar eclipse*, while the latter is a *lunar eclipse*.

If the Moon always moved in the ecliptic plane, then during each new moon it would be positioned on a straight line that connects the Sun and the Earth. Solar and lunar eclipses would then occur each month. In fact, as we know, the moon's orbit is titled with respect to the ecliptic plane. Because of this, when there is a new moon and a full moon, the Moon usually appears either above or below the ecliptic plane. When this happens, there is no way that an eclipse can occur.

It stands to reason that we can see lunar eclipses much more often than solar eclipses. After all, if the Earth's shadow covers the Moon's disk, then this can be seen by everyone on the Earth to whom the Moon is visible at that moment. During a solar eclipse the shadow of the Moon slides across the surface of the Earth in a very narrow band. This is the exact spot where one sees a total solar eclipse (Fig. 2.9).

If a person does not specifically get ready to see a partial solar eclipse, it is very easy to miss it.

Let's examine these two occurrences more closely.

Solar eclipses occur when the Moon partially or completely covers the Sun from the view of an observer on the Earth. A solar eclipse is possible only

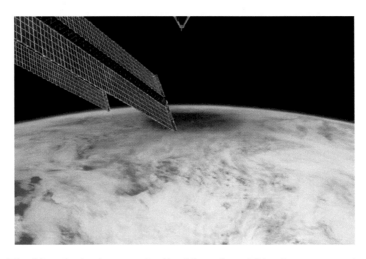

Fig. 2.9 The Moon's shadow on the Earth's surface. This picture was taken by the International Space Station

during a new moon when the Earth, the Moon and the Sun are on the same straight line with the Moon positioned precisely between the Earth and the Sun (Fig. 2.10).

The special characteristics of solar eclipses are visible to us on the Earth due to the fact that the angular diameter of the Moon and the Sun are very close in value. The diameter of the Moon varies from 29′ 3″ to 33′ 5″, while the diameter of the Sun varies from 31′ 3″ to 32′ 3″. When the angular diameter of the Moon is equal to or greater than that of the Sun, a total eclipse occurs; when it is less, an annular solar eclipse is observed. Since the Moon's shadow on the Earth's surface does not exceed 270 km (167.77 mi) in diameter, during a solar eclipse its shadow appears as a narrow band that travels across the Earth's surface. That being said, it is fairly difficult to see this phenomenon from any one specific place. It is necessary to be sufficiently prepared and use caution when looking at partial and annular solar eclipses. This is because the visible part of the Sun continues to shine very brightly and if you look at it—especially through a telescope—without proper eye protection, you can damage your retina. This is why people used to look at eclipses through smoked glass and exposed film, but today we can use light filters to see them.

This is how a total solar eclipse appears if you look at it from a spot on the Earth's surface. The Moon "crawls over" the Sun and slowly covers it

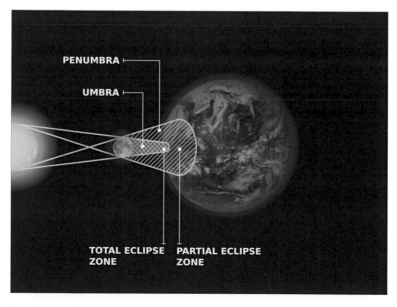

Fig. 2.10 The conditions that must be met for a solar eclipse to occur

(Fig. 2.11). The rotation of the Earth around its axis results in both the Sun and the Moon moving across the sky.

Lunar eclipses occur when the Earth is between the Sun and the Moon and the Moon is in a shadow cast by the Earth (Fig. 2.12), i.e., at the full moon phase.

The Earth's shadow is shaped like a cone and it is easy to calculate— from the Earth to the Moon the diameter of this cone is almost 2.5 times

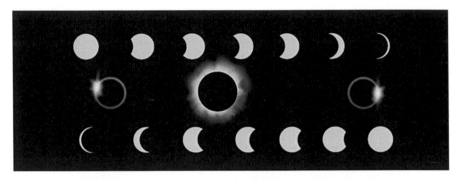

Fig. 2.11 The phases of a solar eclipse as seen from a spot on the Earth's surface

Fig. 2.12 The conditions that must be met for a lunar eclipse to occur

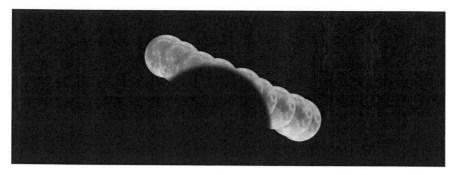

Fig. 2.13 Different phases of a lunar eclipse superimposed onto a picture

greater than that of the Moon's diameter. As a result, when a lunar eclipse occurs, the Moon can be completely hidden. After seeing lunar eclipses, Greek astronomers concluded that the size of the Moon is much smaller than that of the Earth. This is because at certain times the Moon is completely in the Earth's shadow, which is evident if we superimpose images of different phases of a lunar eclipse onto a picture (Fig. 2.13).

Lunar eclipses occur more seldom than solar eclipses, but they are easier to see with the naked eye. Thus, lunar eclipses are seen on half of the Earth's surface, while partial solar eclipses can be observed on a fourth of it. However, total solar eclipses are visible only in a band that is about 250 km (155.34 mi) wide. For this reason, solar eclipses are seldom seen in the same place—this happens about once every 300 years.

Nevertheless, there are years when seven eclipses occur—four solar eclipses and three lunar eclipses—which is the maximum number possible. In ancient times, it was already clear to people that after 18 years and 10 days eclipses recur. Therefore, if there were, for example, seven eclipses in a certain year (naturally they would not all be visible in the same place on the Earth) after 18 years, there will again be a year in which seven eclipses would occur. Moreover, each of them will be 10 days later than they had been 18 years ago.

It used to be that astrologers and different types of fortune tellers thought that solar and lunar eclipses were omens and signs from higher powers. Sometimes these phenomena were received favorably, while other times they were considered warnings of impending doom. Today, however, when the true reasons behind these fascinating occurrences are known, one can confidently assert that there is nothing supernatural about them. After all, when a moth blocks the light of a lightbulb, we do not think it is a bad omen.

2.3   High and Low Tides

The physical nature of the tidal effect lies in the non-uniformity of the gravitational field in space. If a body of finite size enters this field, then different parts of it will experience different levels of acceleration. For a gaseous fluid or liquid, this leads to the displacement of some parts of this body with respect to others; for a solid, this displacement causes subsurface stress to occur. The rising and falling of the Earth's sea level is a result of the Earth moving in a non-uniform gravitation field created by the Moon and the Sun.

Because of gravitational pull, interacting bodies become deformed. If we imagine them as round and liquid, then they will begin to "stretch" and their radius will increase along the semi-major axis by the magnitude ΔR as compared to the radius R of a round body.

Table 2.1 shows the relative changes in the shape of several interacting celestial bodies. Based on this information, one sees that bodies with smaller mass "suffer" from tides more than bodies with larger mass. Hence, of all the bodies shown in Table 2.1, Jupiter's satellite Io is affected the most by deformation. Sometimes a small cosmic body, which was pulled into a large body's low earth orbit, is torn to pieces by tidal forces and rotates around the large body in a circle made up of a tremendous amount of debris. According to one explanation, this is how Saturn acquired its rings.

Deformation of Bodies Due to Gravitational Pull Let us assume that a body with the mass m, which creates a gravitational field around itself, is located at point A (Fig. 2.14). At point B, which is located at the distance r from point A, the acceleration will be equal to:

$$\mathbf{a}(\mathbf{r}) = \frac{Gm}{r^3}\mathbf{r}. \tag{2.4}$$

Now we will find the difference in acceleration at the points that are located at the distance ΔR from point B, which is relative to the rate of acceleration at point B. The difference in

Table 2.1   The changes in the shape of interacted celestial bodies

Causative body	Disturbed body	m/M	R/r	Relative change in shape
Earth	Moon	81.3	4.5×10^{-3}	1.5×10^{-5}
Moon	Earth	1.2×10^{-2}	1.7×10^{-2}	1.2×10^{-7}
Sun	Venus	4.1×10^{5}	5.6×10^{-5}	1.4×10^{-7}
Venus	Sun	2.4×10^{-6}	6.5×10^{-3}	1.3×10^{-13}
Jupiter	Io	2.6×10^{4}	4.3×10^{-3}	4×10^{-3}
Io	Jupiter	3.8×10^{-5}	0.17	4×10^{-7}

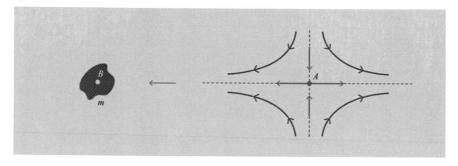

Fig. 2.14 Effect of gravitation on the planet's shape

the rate of acceleration $\Delta a = a(\mathbf{r} + \Delta \mathbf{r}) - a(\mathbf{r})$ will be equal to

$$\Delta a = \frac{Gm}{r^3} \Delta r \qquad (2.5)$$

if the points lie on a plane that is perpendicular to the straight line AB and if the difference in the rate of acceleration is channeled perpendicularly to this line. If the points lie on the straight line AB, then the difference in the rate of acceleration is

$$\Delta a = \frac{2Gm}{r^3} \Delta r \qquad (2.6)$$

and is channeled along the AB line. The directions of the acceleration difference of the vector are shown in Fig. 2.14.

Thus, if a body of finite dimensions is moved to point B, then the tidal force will tend to compress it in the directions that are perpendicular to the axis—which connects the body to the mass that has created a gravitational field—and stretch it along the line that connects this body to the field's sources. If a celestial body located in the gravitational field of another body is easily deformed, then its shape will change. A gaseous envelope (the atmosphere) or a liquid envelope (the hydrosphere) is naturally deformed much more easily than the solid nucleus of a celestial body. When this type of deformation occurs, the body takes the shape of an elongated ellipsoid and the gaseous or liquid envelope takes the shape of an atypical tidal bulge.

The degree to which the shape of a celestial body is distorted, as well as the standard height of tidal bulges, can be estimated as the ratio of tidal acceleration $a_{\text{high tide}} = \Delta a$ to the acceleration of the force of gravity on the surface of a disturbed body. It is it equal to:

$$\frac{\Delta R}{R} = 2 \frac{m}{M} \left(\frac{R}{r} \right)^3. \qquad (2.7)$$

Here we have designated the mass and radius of the disturbed body as M and R; m is the mass of the disturbed body, while r is the distance between the causative and the disturbed bodies.

Let's consider the "planet-satellite" system. If the rotation of a planet or satellite around its axis does not occur synchronously with its orbital revolution, then not only will its shape be distorted, but other tidal effects will also occur. Tidal friction and spin-axis precession are the most important of them. If the planet's period of diurnal rotation is shorter than the satellite's orbital period, then viscosity will cause tidal bulges to be ahead of the satellite as shown in Fig. 2.15.

Since the force of gravity pulling the satellite toward the neighboring bulge is stronger than the pull to the distant satellite, the orbital motion of the satellite will be accelerated and it will begin to move away from the planet. At the same time, the planet's spin rate around its axis will decrease. In the "planet-satellite" system, the total momentum remains. Thus, the effect of tidal friction will be contained in the redistribution of momentum between celestial bodies.

Now let's turn our attention to our planet. The Earth experiences tidal agitation from the Moon and the Sun; however, the Moon's influence is roughly twice as strong. Let's first focus on high and low tide, the occurrence of which is related to a slight distortion of the Earth's shape due to gravity from the Moon. When the Earth rotates, each point on its surface gets closer to the Moon with the period

$$T_{\text{high tide}} = \left(\frac{1}{T_{\text{E}-\text{S}}} + \frac{1}{T_{\text{M}-\text{E}}} \right)^{-1} = 24\,\text{h}\,50\,\text{min}. \tag{2.8}$$

However, as we well know, the interval between high tides is exactly twice as less, i.e., 12 h and 25 min. How can this be explained? The gravitational pull between the Earth and the Moon results in the Earth's surface having two tidal maxima, which run along the Earth's surface as they bend around

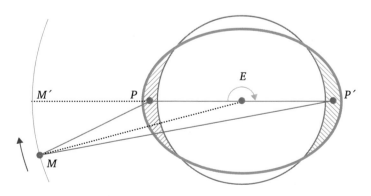

Fig. 2.15 An illustration of tidal friction

it. It is known that the Earth revolves around the center of the "Earth–Moon" system at the accelerated rate $a = \frac{GM_M}{R_{E-M}^2}$. Let us assume that at point A on the Earth's surface, there is a body of the mass m, which is as close to the Moon as possible. Then the force of gravity $F_1 = \frac{GM_Mm}{(R_{E-M}-R_E)^2}$ (Fig. 2.16), which is coming from the Moon, will act on this body. If that same body is close to the Earth, which is as distant from the Moon as possible at point B, then the respective force will be equal to $F_2 = \frac{GM_Mm}{(R_{E-M}+R_E)^2}$, while the average force coming from the Moon, $F_{avg} = \frac{GM_Mm}{R_{E-M}^2}$, as one would guess, is between these values.

Therefore, at an average rate the body at point A accelerates faster than the Earth does, while at point B the body accelerates more slowly. We see that the tidal maximum at point A occurs due to stronger gravitational pull on the Earth's surface with respect to its average amount, while the opposite happens to the tidal maximum at point B due to a weakening of the force of gravity with respect to its average amount.

A more accurate calculation gives the maximum tidal acceleration in the following formula:

$$a_{high\ tide} = \frac{3}{2}\frac{GM_M R_E}{R_{E-M}^3}. \tag{2.9}$$

Tidal acceleration occurs as is shown in Fig. 2.17 where the numbers show a specific value of tidal strength.

Figure 2.17 shows that the intensity of the tidal force is at its maximum at mid-latitudes. This causes tidal streams in the oceans, which can reach a speed of 10 km (6.21 mi)/h. The Earth's terrain is also affected by tidal forces; it rises and falls 30–40 cm (11.81–15.75 in), but does not shift horizontally. Imbalances in geological strata, which lead to fluctuation in the water table,

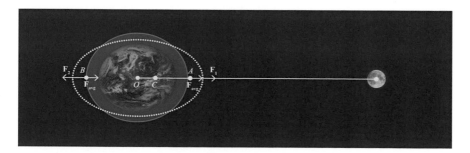

Fig. 2.16 High tides in the "Earth–Moon" system

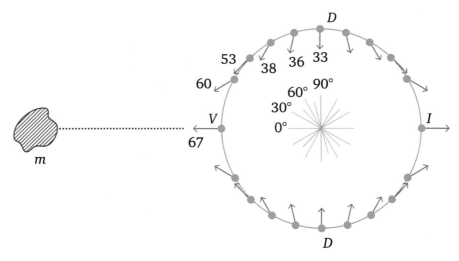

Fig. 2.17 The direction and magnitude of tidal force (in relative units) created by a satellite with the mass *m* at different points on the Earth's surface

changes in oil levels and even sudden gas emissions from coal mines, are all linked to the influence of tidal forces on the Earth. Thus, understanding tides and how they act is important not only for sailors, but also for miners.

We can get an estimate of how distorted the shape of the Earth is from its interaction with the Moon:

$$\Delta R_{\mathrm{E}} \approx R_{\mathrm{E}} \frac{a_{\mathrm{high\ tide}}}{g} \approx R_{\mathrm{E}} \frac{M_{\mathrm{M}}}{M_{\mathrm{E}}} \left(\frac{R_{\mathrm{E}}}{R_{\mathrm{E-M}}} \right)^3 = 0.36\,\mathrm{m}\ (14.17\,\mathrm{in}). \quad (2.10)$$

This estimate is true with regard to the Earth's crust. Tides actually reach a much greater height: they can be about half a meter (about one foot) in the open ocean and much higher near the coast especially in narrow bays and shallow seas (Fig. 2.18). It is easy to understand the reason for this discrepancy. It is due to the fact that since water is a liquid, it changes its shape much more "willingly" than the Earth's crust when it comes in contact with external forces.

Not only the Moon influences the phenomena that take place on the Earth, but the Sun does as well. We will estimate the relative height of tides due to the Sun:

$$\Delta R_{\mathrm{E}} \approx R_{\mathrm{E}} \frac{M_{\mathrm{S}}}{M_{\mathrm{E}}} \left(\frac{R_3}{R_{\mathrm{E-S}}} \right)^3 = 0.16\,\mathrm{m}\ (6.30\,\mathrm{in}). \quad (2.11)$$

Fig. 2.18 The northwest coast of France: **a** at low tide, **b** at high tide

As was already indicated, the Sun's influence on tides is approximately two times less than the Moon's influence on them.

Tides on the Earth reach their maximum height when the Moon and the Sun act together, which is during new moon and full moon when the Earth, the Moon and the Sun are on one straight line (Fig. 2.19a). Consequently, during new moon and full moon high tide is strong and at its highest level, while low tide is at its lowest level.

During the Moon's other phases that occur between new moon and full moon in the first and last quarters, high tide is weak and the difference in water level at high and low tide decreases by more than half compared to what it is when high tide is strong.

One more factor influences the formation of tides. When the Moon is closest to the Earth at perigee its tidal force is greater than when it is farthest from the Earth at apogee. This difference in distance provides a 30% change

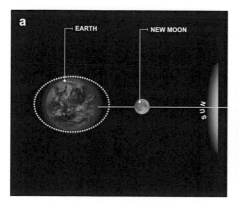
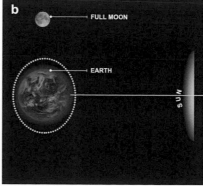

Fig. 2.19 The location of the Sun, the Earth and the Moon: **a** at maximum tidal range, **b** at minimum tidal range

in the height of lunar tides. Changes in the range of lunar tides, along with changes that occur from the combined effect of lunar and solar tides, all result in very significant variations in the height of ocean tides.

To sum up, we can say that the highest tides occur twice a month during new and full moon, which is when the lunar and solar tides combine forces. In the intervals between these very high tides, which are in the first and last quarters, the smallest tides occur when the solar and lunar tides act against each other. New moon usually occurs once a year at the time that the Moon is near perigee; six months later full moon occurs near perigee. The highest tides that occur during these two periods are especially high because the strength of lunar tides significantly increases. These higher-than-average high tides are more than a month late every year because the line that connects the perigee and apogee of the lunar orbit rotates in the same direction as the Moon itself.

Besides existing in oceans, high and low tides are also found in the Earth's gaseous envelope, that is, in its atmosphere. Tidal forces cause periodic fluctuations in atmospheric pressure at the Earth's surface—about 0.22 mm (0.008661417 in) Hg. But the strength of these fluctuations is much smaller than the pressure fluctuations that are associated with changes in the weather. For this reason, they do not affect people's daily lives.

2.4   Precession of the Earth's Axis

As we have already stated, precession is the slow movement of the Earth's axis around the ecliptic pole (Fig. 2.20). The angle of rotation within one year is 20 s and after 26,000 years the axis will circumscribe the shape of a conical surface around the ecliptic pole and return to its original position.

The Earth's interaction with the Moon is essentially the main factor that explains why precession developed in the Earth's orbit. It was Newton who came to this conclusion. The physical reason for precession is the Earth's oblateness at the poles. Although the ratio of the polar and equatorial radii is small (i.e., about 1/300), this deviation in the Earth's shape from that of a sphere creates a moment of force that tends to turn it in such a way so that the equatorial and ecliptic planes coincide. The Moon's orbit is influenced by the Sun and the planets and because of this, over the course of dozens of years the lunar orbital plane and the direction to perihelion only change slightly. As shown in Fig. 2.21, many moon orbits are situated inside of the torus for a long period of time. Although the equatorial bulge of the Earth is small, it is enough for the Moon to cause a moment of force that acts on the Earth.

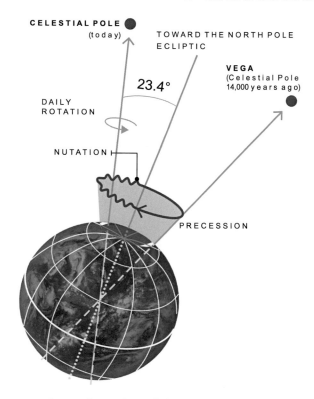

Fig. 2.20 The precession and nutation of the Earth's axis

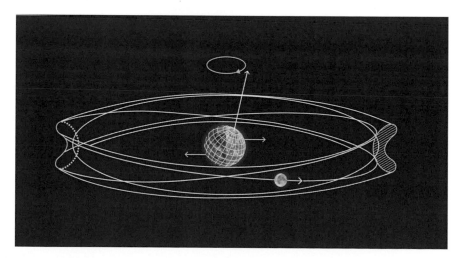

Fig. 2.21 Possible lunar orbits and forces that cause the precession of the Earth's axis

The precession of our planet's axis results in several important ramifications. First, because of precession, the duration of the calendar year—the time during which a complete change of seasons occurs—is shortened. This calendar year has proven to be 20 min shorter than the time the Earth revolves around the Sun.

However, this is not the only example. The Celestial Pole, which is the point of the celestial sphere around which stars' diurnal rotation, which is visible from the Earth, occurs, now coincides with the North Star; in other words, it is located within the constellation Ursa Minor (Fig. 2.22). After approximately 100 years it will begin to move from the Celestial Pole so that after three millennia it will be in the constellation Cepheus. In the days of ancient Babylon, the Celestial Pole was in the constellation Draco.

Due to the fact that the calendar year and the period of the Earth's revolution around the Sun do not coincide, the equinoxes occur a bit earlier than before the Earth completes its full revolution in orbit. That is why precession is also called the *precession of the equinoxes*. Therefore, because of precession, the appearance of constellations and individual stars is gradually changing. For example, if we in the Northern Hemisphere are able to see the constellations

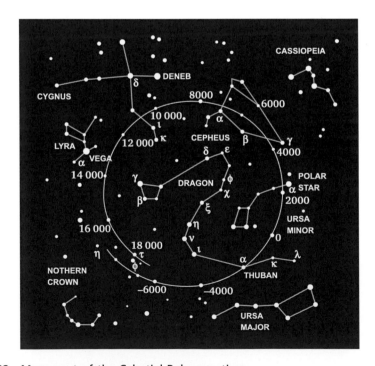

Fig. 2.22 Movement of the Celestial Pole over time

Canis Major and Orion in our lifetime, after 4000 years these constellations will disappear beyond the horizon. However, such constellations as the Southern Cross and Centaurus will become visible to us. Additionally, 6000 years from now people who live in the Southern Hemisphere will have the opportunity to see Ursa Major.

Not only does precession affect the Earth, but its interaction with the Moon results in another type of distortion of the Earth's orbit, which is called *nutation*. It occurs because of the fact that the lunar orbital plane does not coincide with the Earth's orbital plane—the angle between them is 5°. The increase in nutation is small—its range does not exceed 9.2″ and its period is about 18.6 years. Because the amount of nutation that is superimposed on precession is small, the cone circumscribed by the Earth's axis becomes sinuous (see Fig. 2.20).

These clarifications illustrate that physical models, just like Russian nesting dolls, complement each other: each subsequent adjustment makes our projections more accurate, but at the same time our calculations become more complicated. It is necessary to constantly monitor all of these complex motions of the Earth's axis so that we can adjust navigation. This is done by the International Earth Rotation and Reference Systems Service.

2.5 Deceleration of the Earth's Rotation

Another important consequence of the Moon's influence, which was already mentioned, is the deceleration of the Earth's rotation around its axis. Tidal friction leads to 5×10^{12} of heat being released every second, which is about 20% of the terrestrial heat flow. Since the Earth moves with greater angular velocity than the Moon orbiting around it, the tidal maxima that act on the Earth are not exactly on the "Earth–Moon" line, but are turned approximately 2° toward the direction of the Earth's rotation. This shift of tidal maxima creates a moment of force that slows down the Earth's rotation (Fig. 2.23). As a result, the angular velocity with which the Earth rotates around its axis decreases every year by 2×10^{-10} of its magnitude and the duration of a day on the Earth increases every year by 2×10^{-5} s. An outcome of this same phenomenon is that each year the Moon moves approximately 3.8 cm (1.50 in) away from the Earth. Estimates based on the study of fossils show, for example, that 400 million years ago days were 2 h shorter than they are today.

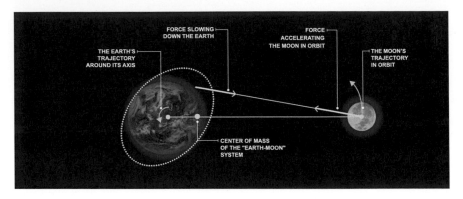

Fig. 2.23 Forces slowing down the Earth's movement and accelerating the Moon's movement

The Earth rotates around its axis faster than the "Earth–Moon" system rotates around a general center of mass. Tidal energy is converted into heat from friction when water moves and from fluid friction against the ocean floor. Hence, we have a deceleration of both the Earth's rotation rate around its axis and the Moon's rotation rate around the Earth. At the same time, the energy from the Moon's rotation around the Earth increases, which leads to an increase in the average distance from the Earth to the Moon.

In the early stages of the evolution of the Solar System, the interaction between the Earth and the Moon led to a deceleration of the Moon's rotation around its axis. As a result of strong tidal forces, the Moon now faces the Earth on one side and the center of its mass has shifted about 3 km toward the Earth in relation to its geometric center.

The deformation of the shape of the Moon can be estimated by using the formula we obtained earlier:

$$\Delta R_{\mathrm{M}} \approx R_{\mathrm{M}} \frac{M_{\mathrm{M}}}{R_{\mathrm{E}}} \left(\frac{R_{\mathrm{M}}}{R_{\mathrm{E-M}}} \right)^3 = 13 \, \mathrm{m} \, (42.65 \, \mathrm{ft}). \tag{2.12}$$

The fact that its actual distortion is much larger once again confirms that long ago the Moon was a great deal closer to the Earth than it is today. About two billion years ago, the Moon was only three earth radii from us; at that time tides were several kilometers (several miles) high and caused tremendous destruction.

How will the "Earth–Moon" system keep on evolving? In the future, the Moon will move even farther away from the Earth. Estimates suggest that five billion years from now the radius of the lunar orbit will reach its maximum, which will be about 460,000 km (285.83 mi). At that time the Earth will stop decelerating because the orbital period around its axis will be equal to the orbital period of the Moon around the Earth. Thereafter, the Earth will always face the Moon on one side. Tidal friction will not disappear because solar tides will still slow down the Earth's movement, but this deceleration and tidal range will not exist as much as it does now.

How can one most precisely find out what the length of days was in the distant past? Paleontology will allow us to do this. Scientists discovered fossil colonies of blue-green algae, which were hundreds of millions of years old. These colonies were similar to modern-day coral; in particular, they grew much faster in the daytime than at night. Consequently, diurnal rings similar to annual rings on the cross sections of trees can be found on the cross sections of these algae. About 200 million years ago, a year consisted of 385 days; 400 million years ago, a year was made up of 400 days; and 600 million years ago, a year had 425 days. Thus, paleontological data prove that 600 million years ago, a 24-h period consisted of only 20 h, which confirms the validity of astronomers' calculations.

Further Reading

1 Aderin-Pocock, M.: The Book of the Moon: A Guide to Our Closest Neighbor. Abrams Image (2019)
2 Byalko, A.V.: Our Planet the Earth. MIR Publisher (1983)
3 Close, F.: Eclipses: What Everyone Needs to Know. Oxford University Press (2019)
4 Hockey, T.A.: The Book of the Moon: A Lunar Introduction to Astronomy, Geology, Space Physics, and Space Travel. Simon & Shuster (1986)
5 Kopal, Z.: Physics and Astronomy of the Moon. Academic Press (2017)
6 Varlamov, A.A., Aslamazov, L.G.: The Wonders of Physics, 4th edn. World Scientific (2019)

3

The Depths and the Surface of the Earth

Abstract In the third chapter, we discuss what is the equilibrium condition of the planet's surface in its own gravitational field and find out the planet's shape deformation and the maximum height of a mountain on a particular planet. Then we consider the physical effects related to meteors' collision with the Earth's surface, with the main emphasis on the formation of craters. The Earth's internal structure, the movement of the Earth's crust and the dynamics of tectonic plates are then discussed, as well as such phenomena as volcanoes and geysers. The physics of earthquakes and seismic waves is reviewed. In addition, we talk about the geographic coordinate system and the physical background of satellite navigation.

One of the most stunning images seen by a person who has traveled to the Moon is the Earth in the night sky (Fig. 3.1). There are millions and even billions of similar objects in the Universe. Except for the Earth. It is, after all, the only planet that as of today is known to have intelligent life forms on it. And it is precisely because the Earth is so unique that it poses such a great challenge to modern-day science.

Today much more is known about the structure of stars that are thousands of light years away from us and the phenomena that occur deep within them than what happens right under our own feet in the bowels of the Earth just a few thousand kilometers (hundred miles) away from us. Therefore, it is very important to study the Earth by relying on different fields of science because

© The Author(s), under exclusive license to Springer Nature
Switzerland AG 2023
D. Livanov, *The Physics of Planet Earth and Its Natural Wonders*,
https://doi.org/10.1007/978-3-031-33426-9_3

Fig. 3.1 The first picture of the Earth taken on August 23, 1966 during the flight of the Lunar Orbiter 1 spacecraft

this planet holds many secrets and mysteries that have not only scientific but also practical significance. The Earth not only gives people opportunities to live and develop, but it often poses serious threats to people's safety: earthquakes, floods, volcanic eruptions and destructive ocean waves can all put people's lives in jeopardy. In addition, man-made technology and the outcome of people's actions often subject not only individuals to danger, but the Earth and all of humanity as a whole as well. We are all faced with the challenge of saving the Earth, our beautiful home. In order to do this, we must first understand why our planet is the way it is.

3.1   The Shape of Planets

Why is the shape of all planets, including the Earth, almost round? The reason for this is because of the law of universal gravitation. The larger a planet's mass, the more all of the bodies on a surface are attracted to it. Because of gravity, the protruding distortions on planets break off and their shape becomes smooth.

If a cosmic body has a radius greater than $R_{max} \approx \frac{p_{max}^{\frac{1}{2}}}{\rho G^{\frac{1}{2}}}$ where p_{max} is the pressure created in the central region of a celestial body, ρ is the density of matter and $G = 6.67 \times 10^{-11} \frac{m^3}{s^2 \, kg}$ is the gravitational constant, then this cosmic body will have a spherical shape. Small bodies, however, that are $R < R_{max}$. in size will retain an irregular shape since the force of their own gravity is not enough to flatten their surface and make them round.

In Table 3.1, we will provide the calculated values of R_{max}. for the substances of which celestial bodies are composed.

The size of all of the planets in the Solar System is much larger than a critical dimension, which is why they are almost round. However, Mars' satellite Phobos (Fig. 3.2a), for example, is shaped like a potato because its size is 14 × 11 × 10 km (8.70 × 6.84 × 6.21 mi), which obviously does not reach the significance point.

The asteroid Itokawa is approximately 500 × 300 × 200 m (1640.42 × 984.25 × 656.17 ft) in size and resembles a large, bumpy cucumber (Fig. 3.2b).

But to say that the Earth is absolutely round is another idealized approximation. Indeed, as was already stated, due to the rotation around its axis it is flattened at the poles—its polar radius R_{pole} is less than its equatorial radius R_{eq}. The ratio of the difference between the equatorial and polar radii to the equatorial radius is only one three-hundredth.

In actuality the surface of the planets is far from being completely round; valleys and mountains make their geographical landscape much more diverse. The closer you get to a planet, the greater the number of different types of uneven surfaces you will see. When a person is somewhere in the Himalayas

Table 3.1 Relevant physical parameters for substances of which celestial bodies are composed

	Ice	Moon rocks or Martian meteorites	Granite	Iron
ρ, 10^3 kg/m^3	1.0	2.5	7.0	7.8
p_{max}, N/m^2	3×10^6	3×10^7	3×10^{20}	3×10^{21}
R_{max}, km	200	300	500	500

 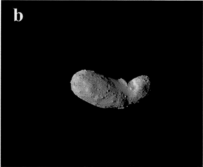

Fig. 3.2 Irregularly shaped celestial bodies: **a** Mars' satellite Phobos, **b** the asteroid Itokawa

and looks at their majestic snow-capped peaks, it is difficult to believe that from space everything seems to be one part of a smooth ball.

Estimating the height of mountains on a particular planet is quite simple. Let's assume that a mountain is shaped like a cone and has the height h. Then the average pressure at the base of the cone will be $p = \frac{1}{3}\rho g h$. A mountain can exist as long as $p < p_{max}$, hence the maximum height of mountains on the Earth is:

$$h_{max} \approx \frac{3p_{max}}{\rho g} \approx 11 \text{ km (6.84 mi)}. \qquad (3.1)$$

As we are aware, the maximum height of mountains on the Earth is about 9 km (5.60 mi); thus, our rough estimate is entirely plausible despite the fact that mountains with a shape as perfect as a cone do not exist. Nevertheless, one can find several examples of mountains that are indeed perfectly shaped (Fig. 3.3).

The Maximum Radius at Which a Body Can Have an Aspherical Shape We will determine the critical size that a celestial body must have in order to become shaped like a sphere because of gravity. Let's consider a celestial body made up of solid matter. We will assume that the mass of this body is m and the density is ρ. Then its average linear dimension is $R \approx \left(\frac{m}{\rho}\right)^{\frac{1}{3}}$. The acceleration of gravity on the body's surface produces $g = \frac{Gm}{R^2} \approx Gm^{\frac{1}{3}}\rho^{\frac{2}{3}}$. The pressure created by gravity in the center of this celestial body is $p \approx \rho g R \approx Gm^{\frac{2}{3}}\rho^{\frac{4}{3}}$. When a solid is compressed, it tends to maintain its shape and resists external force. In this way, solids differ from liquids and gases, which have a shape that is much easier to transform. However, a solid's

Fig. 3.3 Cone-shaped mountains: **a** Ama Dablam Mountain, **b** K2

shape will remain the same under pressure only to the defined limit that is characteristic for each particular type of matter. If the amount of pressure exceeds this limit, which is called the *ultimate strength* (p_{max}), then the body ruptures by cracking or crumbling, for example. Thus, this disruptive condition is written as $p > p_{max}$. It follows from this that we can determine the characteristic size of a cosmic body with protrusions that will be ruptured with this formula:

$$R_{max} \approx \frac{p_{max}^{\frac{1}{2}}}{\rho G^{\frac{1}{2}}} \tag{3.2}$$

Relative Deformation of the Earth's Shape An important factor that affects a planet's shape is its rotation around its axis. If its radius is larger than the critical radius, then its equilibrium shape will be oblate at the poles, specifically, the points through which the axis of rotation passes.

A planet's surface at any point must be equal to its resultant force—which is the sum total of gravitational and centripetal forces—in order for the shape of the planet to be in equilibrium (Fig. 3.4).

Let's estimate the relative deformation of a planet with the size R. The force acting at the pole point is entirely determined by gravity; the corresponding acceleration is $a_{pole} = \frac{MG}{R^2}$. At the Equator, however, this force decreases due to the centrifugal force, which leads to a decrease in the total acceleration: $a_{eq} = \frac{MG}{R^2} - \omega^2 R$. The condition that must be met in order for there to be equilibrium on a planet's surface is that at any point the sum total of the vectors of gravitational and centrifugal acceleration is perpendicular to the surface. Then:

$$\frac{R_{eq} - R_{pole}}{R_{eq}} \approx \frac{a_{eq} - a_{pole}}{a_{eq}} = \frac{\omega^2 R^3}{MG} \approx \frac{1}{300}. \tag{3.3}$$

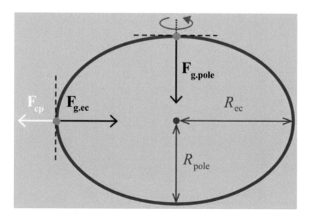

Fig. 3.4 Condition of equilibrium for the planet's form

This estimate approximately corresponds to the measured value of the oblateness of the Earth.

Similar estimates can be made for other planets in the Solar System. It intuitively makes sense that the oblateness of the giant planets—Jupiter and Saturn—which consist mostly of gases and do not have a solid surface, is much larger by comparison (i.e., about 1/10) than that of the Earth, Venus and Mars.

On the Earth's surface there are not only mountains and valleys that formed during the evolution of our planet. Just as with any celestial body one can find traces of cosmic "traffic accidents" on the Earth's surface that were caused by craters.

Craters are found on a planet's surface as a result of collisions with extraterrestrial objects and meteors.

The Meteor Crater in Arizona (U.S.) (Fig. 3.5) appeared on the Earth about 50,000 years ago after a meteorite that was approximately 50 m (164.04 ft) wide and 300,000 T (330,693 sh. tn.) landed there. The explosion from the fall was three times more powerful than the explosion that occurred during the Tunguska Event.[1] A huge bowl-shaped depression was formed with a diameter of more than 1200 m (3937 ft) and a depth of 230 m (754.59 ft). The edge of the crater rises 46 m (150.92 ft) above the plain.

In an impact crater in India a 1800-m-wide (5905.15 ft) and 132-m-deep (433.07 ft) lake appeared (Fig. 3.6), which formed at approximately the same time as the one in Arizona.

Craters have even been discovered on the Moon (Fig. 3.7) and on planets in the Solar System (Fig. 3.8).

The Size of Craters on the Earth We will explain one of the ways to estimate the size of an impact crater that forms on the Earth when a meteorite with the mass m falls at the rate v. If the mass density of which the meteorite consists is ρ, then the transverse size of the meteorite will be $r_m \approx \left(\frac{m}{\rho}\right)^{\frac{1}{3}}$.

One would like to think that the size of the impact crater and of the fallen meteorite will be equal. This would indeed be the case if the rate of fall v were small and less than the

---

[1] This was a massive explosion that was attributed to the air burst of a stony meteorite about 100 m (330 ft) in size near the Podkamyennaya Tunguska River in present-day Krasnoyarskiy Kray, Russia, on June 30, 1908 (translator's note).

Fig. 3.5 Meteor Crater National Landmark (Arizona, U.S.)

Fig. 3.6 Lonar Lake formed from the impact of a meteorite

speed of sound in the corresponding substance (terrestrial rock) $v_{sound} = 5 \times 10^3$ m/s. Since the speed of the falling meteorite is $v > v_{sound}$, the crater will be much larger.

Suppose that the speed of a meteorite entering the Earth's atmosphere is approximately equal to the second cosmic speed 11.2 km/s. Given that the meteorite is slowing down in the Earth's atmosphere, let us assume that $v = 10^4$ m/s. The kinetic energy of the falling meteorite is $E = \frac{mv^2}{2}$. On what is this energy spent? First, on the formation of a depression (i.e., an impact crater) in the Earth's surface due to the destruction and crushing of terrestrial

Fig. 3.7 Daedalus (a crater on the Moon)

Fig. 3.8 Craters on Hellas (Mars)

rocks. Second, on the projection of particles from the crater. We will denote their energy magnitude as E_1 and E_2, respectively.

Let us assume that the size of the resulting impact crater is R_c; then its volume is $V_c \approx R_c^3$. It is necessary to use an amount of energy that is equal to the product of this volume on the ultimate strength of the sedimentary rock $p_{max} \approx 10^7 \, \text{N/m}^2$ in order to break a rock with the volume V_c. Therefore, $E_1 \approx p_{max} R_c^3$. When an impact crater is formed, substance scatters at a distance comparable to the size of the crater itself. In order for substance to be ejected from the crater at the distance R_c, it must have an initial speed of $v_0 \approx \sqrt{g R_c}$. The mass of terrestrial rocks ejected from the crater is $M \approx \rho R_c^3$, where $\rho = 3 \times 10^3 \, \text{kg/cm}^3$, which is the density of terrestrial rock. Thus, the energy required to eject the rock from the crater will be equal to $E_2 \approx M v_0^2 = \rho g R_c^4$. Consequently, when a meteorite falls, the law of conservation of energy is determined by the formula $E = E_1 + E_2 = p_{max} R_c^3 + \rho g R_c^4$. Or

$$\frac{mv^2}{2} = p_{max} R_c^3 + \rho g R_c^4. \tag{3.4}$$

For small craters the first sum on the right side will be dominant, while for large craters the second one will. Two sums are compared when

$$R_{c(0)} \approx \frac{p_{max}}{\rho g} \approx 300 \, \text{m} \, (984.25 \, \text{ft}). \tag{3.5}$$

A crater of this type is formed when a meteorite with the mass $m_{(0)} = 3 \times 10^6 \, \text{kg} = 3000 \, \text{T} (3306 \, \text{sh. tn.})$ falls. But what will happen if a huge meteorite falls into the ocean? The kinetic energy of the falling meteorite $E = \frac{mv^2}{2}$ will be expended on the evaporation of water with the volume $V_c \approx R_c^3$ and a splash of water R_c in height. We will write the law of conservation of energy as:

$$\frac{mv^2}{2} = (\lambda + c\Delta T)\rho R_c^3 + \rho_{water} g R_c^4, \tag{3.6}$$

where $\lambda = 2.5 \times 10^6 \, \text{J/kg}$ is the specific heat of evaporation, $c = 4.2 \times 10^3 \, \frac{J}{\text{kg K}}$ is its specific heat capacity and ΔT is the temperature at which warming occurs.

It is easy to understand that a crater on the ocean's floor is formed if R_c is greater than the depth of the ocean at a given area. If, for example, one considers that the depth is equal to the average depth of the ocean, which is 4 km (2.49 mi), then the mass of the meteorite should exceed $3 \times 10^9 \, \text{T}$.

If such a catastrophic event as this occurred, then a giant tsunami wave would go around the entire world several times and a tremendous cloud of evaporated water would cause a torrential downpour everywhere on the surface of the Earth.

A crater of about 300 m (984.25 ft) in size is formed from the impact of a body weighing 3000 T (3306 sh. tn.).

Meteorites are small bodies of cosmic origin that revolve around the Sun and fall onto a planet's surface.

Celestial bodies that measure up to several meters (feet) are called *meteoroids*, while larger ones are called *asteroids*. Small meteoroids that completely burn out and leave a train in the Earth's atmosphere are called *meteors*, whereas larger ones that explode into a bright flash are called *fireballs*. If the mass of a meteoroid reaches several tons, then it usually collides with the Earth, but prior to this collision its form completely changes because its speed significantly decelerates in the atmosphere. Large meteorites, asteroids or fragments of comets that weigh more than 10 T (11.02 sh. tn.) eventually reach the Earth's surface in a slightly transformed shape. When large meteorites composed of iron or stone make contact with the Earth, they become deformed and an impact crater forms where they fell. The last vivid example of a meteorite that entered the Earth's atmosphere was the Chelyabinsk meteorite (Fig. 3.9). It fell from the sky on February 15, 2013 and, according to estimates, it was 17 m (55.77 ft) in diameter with a mass of about 10,000 T (11,023 sh. tn.). When this meteorite was passing through the thick layers of the atmosphere, it broke into pieces, after which there were bright flashes of light and a change in luminosity. Its sharp deceleration in the atmosphere created a shock wave, which emitted an amount of energy equivalent to an explosion with the strength of about 2×10^{15} J. This is roughly 20 times the strength of the atomic bomb that was dropped on Hiroshima! The largest fragment of the Chelyabinsk meteorite weighed about 500 kg (1102 lbs) and was found on the bottom of Lake Chebarkul (Fig. 3.10).

The "birthplace" of meteorites is usually the asteroid belt, which is a cluster of many cosmic bodies of different sizes between the orbits of Mars and Jupiter. The fragments that result from collisions between these bodies move with great eccentricity along a trajectory and can cross paths with the Earth's orbit

Fig. 3.9 A train left by a meteorite that fell over Chelyabinsk (Russia) on February 15, 2013

Fig. 3.10 A fragment of the Chelyabinsk meteorite, which was found on the bottom of Lake Chebarkul

and become meteorites. Each year about 2000 meteorites fall on the Earth. Scientists have learned that the age of meteorites is between 500 million and 5 billion years (the approximate age of the Solar System) and they consist of the same chemical elements that exist on the Earth. These facts confirm the unity of all that exists in the Universe and the universality of the laws of nature.

To put this in perspective, when the famous Tunguska meteorite crashed on the Earth, the energy that was released from its explosion was about 2×10^{17} J. The height of the explosion that occurred after this meteorite crashed was 5–10 km (3.11–6.21 mi), its speed at the moment of the explosion was about 40 km (24.84 mi)/s and its initial mass was about 10^6 T. The majority of scientists today are inclined to think that the Tunguska meteorite originated from a comet. The nucleus of comets consists of ice, which comes not only from water, but also from carbon dioxide (CO_2) mixed with methane and ammonia. This ice is contaminated with metal substances. The average density of a comet's core is therefore about 1 g/cm^3. Ice, as we know, is very fragile. When a comet's nucleus enters the atmosphere, the substance of which it is made up begins to crumble and evaporate. Due to mechanical and thermal stress, fragmentation occurs at a rapid rate, which further increases the surface area and, hence, the evaporation rate. If initially the comet's nucleus had the radius R, then the reference area where the evaporation occurred would be equal to $4\pi R^2$. When additional fragmentation occurs, this surface area keeps on increasing over and over again. If a body splits into N fragments, then the total surface area of each one is equal to $4\pi R^2 N^{1/3}$. When comets enter the increasingly dense layers of the atmosphere, their matter is heated up by friction. As a result, their rate of evaporation rapidly increases. Estimates suggest that the evaporation rate of comet matter increases so quickly, even catastrophically, that almost all of this matter instantly turns into steam and causes an explosion. The Tunguska meteorite almost did not reach the Earth's surface; particle matter found where it had crashed had a maximum size of about 5×10^{-5} m. But the explosion caused a tremendous blast, which brought down all of the trees within a 2000-km (1242.72 mi) area in a dense, marshy forest in Siberia.

Can a large meteorite change the duration of the Earth's days, that is, can it speed up or slow down the Earth's rotation? The momentum that was transferred to the Earth while the Chelyabinsk meteorite was falling can be calculated using the formula $\Delta v M \approx m v R$. By substituting the meteorite's parameters $m \approx 10^7$ kg, $v \approx 17$ km (10.56 mi)/s, $R \approx 17$ m (55.77 ft) into this formula, we find that $\Delta M \approx 10^{18} \frac{\text{kg m}^2}{\text{s}}$. After speeding up its rotation around the Earth, the meteorite flew in from the east along a practically tangential path. The momentum of the Earth's rotation is $M_E \approx 10^{33} \frac{\text{kg m}^2}{\text{s}}$. Since the duration of a day on the Earth lasts 86,400 s, the variation in the magnitude is only 10^{-11} s, which is invariably less than the accuracy of any instrument that measures time intervals. If we take into account the fact that the Earth is slowing down, even a small additional amount of rotation will be good for us!

3.2   The Acceleration of Gravity on the Earth's Surface

If a body with the mass m is placed on a terrestrial pole, then only the force of gravity, which by the law of universal gravitation is equal to $F_g = G\frac{m M_E}{R_E^2}$, acts upon it. When we apply Newton's second law, we get a formula for the acceleration of gravity with regard to a fixed system:

$$g = G\frac{M_E}{R_E^2}. \tag{3.7}$$

Everyone knows this from physics' class in school. Assuming the mass is constant, we take into account the difference in the radii at the Equator and at the poles and obtain a range of values from 9780 to 9832 m/s², respectively.

Since the Earth rotates around its axis, any body that is not located at one of the poles experiences centripetal acceleration, which in the frame of reference related to the Earth, brings about a change in the observed acceleration of gravity. Here we must take into account the fact that a body's acceleration with relation to the fixed coordinate system g will be equal to the sum total of the acceleration with regard to the moving coordinate system (we will denote it as g') and to the acceleration of the coordinate system itself in relation to the fixed system, $\omega^2 R_E$. We get

$$g' = g - \omega^2 R_E. \tag{3.8}$$

From this it is clear that the value of the acceleration of gravity measured at the Equator is less than at the poles.

We find the value of the acceleration of gravity at any point on the Earth's surface at the latitude φ in the following way. The centrifugal force at the arbitrary point A is equal to $F_{cf} = m\omega^2 R_E \cos\varphi$. We will apply the law of cosines to the shaded area of the triangle in Fig. 3.11.
Then

$$g' = \sqrt{g - \omega^2 R_E \cos^2\varphi \left(2g - \omega^2 R_E\right)} \approx g - \omega^2 R_E \cos^2\varphi. \tag{3.9}$$

The angular velocity of the Earth's rotation is small; therefore, the difference in the acceleration of gravity at different points on the Earth's surface is insignificant. At the Equator $\omega^2 R_E = 3.4 \times 10^{-2}$ m/s², which is a tenth of a

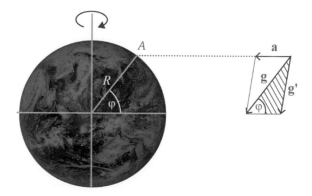

Fig. 3.11 At an arbitrary point on the Earth's surface, the vector of the acceleration of gravity slightly deviates toward the Equator

percent of *g*. It would seem that this is a small correction, but in some cases this difference is very significant.

For example, a clock with a pendulum at the Equator will lose time to the same clock at the poles by more than 3 min in a 24-h period. A rocket launched at the Equator can place heavier objects into orbit than one launched from the polar latitudes even if both have the same amount of propellant. For this reason, launching sites are built in places that are as close to the Equator as possible.

Since the Earth's atmosphere does not have a uniform density, practical experimental measurements of the acceleration of gravity show that it depends not only on latitude, but, in essence, it particularly depends on each point on the Earth's surface. Such measurements allow us to evaluate the actual shape of the Earth and draw conclusions about the internal structure of its depths. Gravimetric analysis studies the ways of measuring the acceleration of gravity in different places and the methods of acquiring information from these data. By speeding up the acceleration of gravity at different places, one can predict the density of rocks. Gravimetric analysis also enables us to refine the geoid, which is a geometric figure that is the same shape as the Earth and is essential for more accurate navigation.

3.3   Coordinates and Navigation

The coordinate system that is used in the world today is very useful for finding any point on a sphere, which is what our planet is. It is called the *geographic coordinate system*. It is two-dimensional, which is natural, since on the surface we need only two values, which are called *latitude* and *longitude*.

> Geographic latitude (φ) is the angle between a line passing through the point perpendicular to the surface of the Earth and the equatorial plane.

Latitude varies from 0 to 90° on either side of the Equator. Northern latitudes lie to the north of the Equator, while southern latitudes lie to the south of it (Fig. 3.12a).

> Geographic longitude (λ) is the angle between two planes that pass through the Earth's axis, one of which passes through the point in question and the

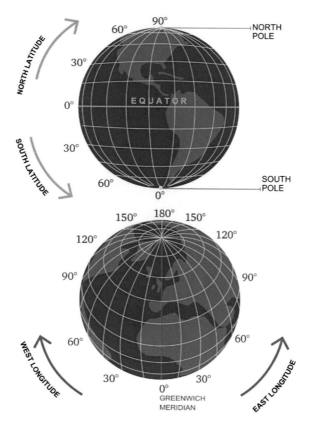

Fig. 3.12 Geographic coordinates

other passes through the Royal Greenwich Observatory, which was selected as the Prime Meridian.

Longitude varies from 0 to 180° to the east and west of the Prime Meridian; when describing the former, it is called *east longitude*, when describing the latter, it is called *west longitude* (Fig. 3.12b).

Let's figure out how the geographical coordinates of a person or a car are determined using satellite navigation systems (e.g., GLONASS or GPS). Such systems contain two components. The satellite (or interstellar) component is a constellation of satellites evenly distributed over the surface of the Earth (Fig. 3.13).

The control segment is an information system located on the Earth, which ensures that the satellites use a single coordinate system and a common time.

Fig. 3.13 GLONASS orbital group diagram

At certain intervals each satellite transmits messages that contain its coordinates at the moment the message was sent and the time at which it was sent. After receiving a message, any receiving device (e.g., a vehicle with a navigation system, gadget or other device), can calculate the distance to the satellite: $l = (t_{rec} - t_{trans})c$ where c is the speed of light, t_{trans} is the signal dispatch time that is found in the satellite transmission and t_{rec} is the signal reception time according to the clock of the receiving device (i.e., navigator). If the satellite coordinates at the time the message is sent are x_{sat}, y_{sat}, z_{sat} and your coordinates are x, y, z, then $l^2 = (x - x_{sat})^2 + (y - y_{sat})^2 + (z - z_{sat})^2$. To determine these three unknown coordinates, three equations are needed. That being said, the receiving device must receive the transmissions simultaneously from three satellites. If this happens, a system made up of three equations with three unknowns emerges:

$$(x - x_{sat1})^2 + (y - y_{sat1})^2 + (z - z_{sat1})^2 = l_1^2; \tag{3.10}$$

$$(x - x_{sat2})^2 + (y - y_{sat2})^2 + (z - z_{sat2})^2 = l_2^2; \tag{3.11}$$

$$(x - x_{sat3})^2 + (y - y_{sat3})^2 + (z - z_{sat3})^2 = l_3^2. \tag{3.12}$$

The geometric interpretation of this system of equations is as follows. The signal from one satellite gives us an aggregate of points, which form a sphere, that are equally spaced from the location of this satellite. The second satellite gives us an aggregate of points that form a second sphere. Both of these spheres then intersect circumferentially. Finally, the third satellite is the third sphere, which intersects with the circumference, thus clearly determining our coordinates (Fig. 3.14). The second root of the square equation (the second intersection point of the circumference and of the third sphere) is rejected because it is physically meaningless.

However, this method of determining the position of a point is not used in the real world because the equation includes the signal reception time according to the navigator's clock. What if this time has been determined incorrectly and the navigator's clock is fast or slow? If the navigator's clock is slow by a microscopic amount, for example, a thousandth of a second compared to the time of the satellite system, this will immediately result in an error of 300 km (186.41 mi) in the measurement of the radius of one of the spheres and make the operation of the entire system invalid.

For all intents and purposes, it turns out that it is not necessary to know the exact time of the navigator. What do we do in this situation? Even if the

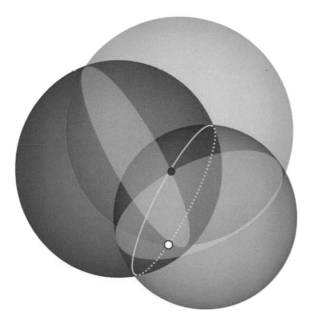

Fig. 3.14 A graphic determination of a point location using information from three satellites

navigator's clock is inaccurate, its display differs from the time on the satellites by the value Δ, which we do not know. In this case, the navigator will incorrectly calculate the distance to the satellite and the discrepancy between this distance and the actual distance will be Δc. Hence, the navigator will identify the distance to the satellite as l, while the actual distance will be $l + \Delta c$.

This means that we now face the challenge of solving a problem with four unknowns: x, y, z and Δ. In order to solve it, four equations are needed; therefore, it is necessary to pick up the signal of another satellite. We get a system of four equations with four unknowns:

$$(x - x_{sat1})^2 + (y - y_{sat1})^2 + (z - z_{sat1})^2 = (l_1 + \Delta c)^2; \qquad (3.13)$$

$$(x - x_{sat2})^2 + (y - y_{sat2})^2 + (z - z_{sat2})^2 = (l_2 + \Delta c)^2; \qquad (3.14)$$

$$(x - x_{sat3})^2 + (y - y_{sat3})^2 + (z - z_{sat3})^2 = (l_3 + \Delta c)^2; \qquad (3.15)$$

$$(x - x_{sat4})^2 + (y - y_{sat4})^2 + (z - z_{sat4})^2 = (l_4 + \Delta c)^2. \qquad (3.16)$$

You can clearly imagine the following scenario. If the navigator's clock is slow, then the receiving device is located in the center of the sphere with the radius Δc, which outwardly comes in contact with four spheres in the center of which are satellites. A simplistic picture is shown in Fig. 3.15 of a two-dimensional variant where one circumference comes in contact with three other circumferences.

After solving this system of four equations, the navigator will tell you not only what your coordinates are, but also fix the error made by its clock and inform you of the "correct" time.

In conclusion, let's consider the question of how to determine the minimum distance between two points on the Earth's surface. Let's see how the flight path from Moscow to a city a significant distance away, for example, Petropavlovsk-Kamchatsky, looks on a world map. Although these two cities are located on approximately the same geographical latitude, the flight path takes a trajectory through much higher latitudes (Fig. 3.16). The airplane's flight path on the map is much longer than the straight line that passes along the line of constant latitude.

Why does this happen? A flat image on a map of the Earth's surface distorts the real shape of the Earth's surface. Additionally, the shortest possible distance between two points is determined by the shape of the surface on

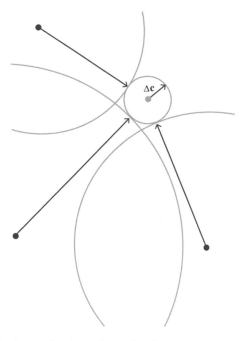

Fig. 3.15 A graphic determination of a point location and time using information from four satellites

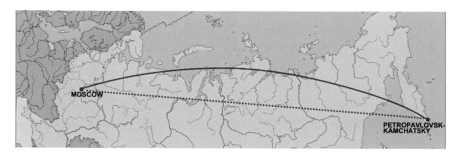

Fig. 3.16 The trajectory of a transcontinental flight (i.e., solid line)

which these points are located. If we assume that the Earth is roughly shaped like a sphere, then the shortest possible distance between the two points of this sphere is determined by the arc of a circumference. It is formed when a sphere intersects with a plane that passes through the designated points and the center of the sphere. There are two such arcs; the shortest possible distance is obviously determined by the length of the smaller one as shown in Fig. 3.17.

In mathematics this type of line that identifies the minimum distance between two points on a given surface is called a *geodetic line*.

Fig. 3.17 A geodetic line

Any other route that connects Moscow and Petropavlovsk-Kamchatsky will be longer than the one that passes through the arc of a circle and connects two cities. Now when you look at a map and see the flight plan of a plane flying a great distance, you won't wonder why it isn't flying in a straight line.

3.4   The Earth's Internal Structure

Now that we have examined the topography and the shape of the Earth's surface, it is necessary to analyze its inner structure.

Studying the inside of the Earth is very difficult. Direct observations of rocks that made their way to the surface and geological examinations including ultra-deep drilling have only revealed information about a very thin layer on the Earth's surface. Suffice it to say that the deepest borehole, which is located on the Kola Peninsula in Russia, has a depth of just over 12 km (7.46 mi), which is only 0.2% of the Earth's radius. Technology has not yet been developed that would make it possible to penetrate into the bowels of the Earth for hundreds and thousands of kilometers (hundreds and thousands of miles) in order to extract samples of geological rocks. Therefore, indirect methods—most notably the seismic method that registers elastic vibrations through the depths of the Earth—are used for geophysical exploration.

Knowing the gravitational constant and measuring the acceleration of gravity made it possible to determine the mass of our planet—6×10^{24} kg—which corresponds to the mean density of matter of the Earth—5.5 g/cm^3. It should be noted that the mass density of the Earth's surface, which is measured directly, is about 2.7 g/cm^3, i.e., about two times lower than average. Based on this information we can conclude that when approaching the central regions of the Earth, mass density increases.

In addition, people have long noticed that heat is constantly flowing from the depths of the Earth. Since heat is transferred from hotter regions to colder ones, we can infer that the temperature deep within the Earth is higher than on its surface. Ultra-deep drilling has proven that the temperature increases by about 30 °C (86 °F) when a drill is submerged 1 km (0.62 mi) into the Earth, although this measurement depends on the location of the drilling point on the Earth's surface.

The Flow of Heat from the Earth's Interior Let's estimate the flow of heat that comes from the depths of the Earth. We will accept the thermal conductivity coefficient of basalt $\kappa = 2 \frac{J}{m\,cK}$ as the thermal conductivity coefficient of the Earth's crust. As we have already mentioned, the Earth's temperature rises with depth. We will add the value $\frac{\Delta T}{\Delta z} = 3 \times 10^{-2}$ K/m, which shows the temperature change during a 1 m (3.28 ft) immersion into the Earth's interior. The heat flux will be equal to the product of the Earth's surface area $4\pi R_E^2$, the thermal conductivity coefficient κ and the temperature variation rate in the given direction $\frac{\Delta T}{\Delta z}$. We have:

$$Q = \kappa\, 4\pi R_E^2 = \frac{\Delta T}{\Delta z} = 3 \times 10^{13}\,\text{W}. \tag{3.17}$$

What is the source of this heat? What is the reason that the bowels of the Earth are hot? The answer lies in our planet's geological history. According to ideas from modern science, the Earth, like other planets in the Solar System, was formed after a large number of small celestial bodies merged together.

While the Earth was rotating around a protostar, it gradually attracted smaller bodies in the early stages of its development, which resulted in an increase in its mass.

Gravitational interaction between any two bodies is governed by universal gravitation and it means that when you divide a body into two parts and separate them from each other at an infinite distance, a certain amount of energy must be spent. However, when two bodies connect because of gravitational pull, energy is released. If there is enough of it, a body will heat up and perhaps even melt when energy is released.

Imagine that we remove a piece of the Earth that has the mass Δm and take it some distance far away. This will require that we use $\frac{G M_E \Delta m}{R_E}$ of energy. The Earth's internal structure, as well as the distribution of mass within it, is well understood. This has enabled scientists to accumulate information about small amounts of mass and accurately calculate the total gravitational energy of the Earth:

$$E_g = -0.4 \frac{G M_E^2}{R_E} = -1.5 \times 10^{32} \, \text{J}. \qquad (3.18)$$

While the Earth was forming, energy was released exactly the same way, but with a positive sign. However, obviously not all of the energy that is stored in the Earth's interior heats it. After all, when bodies collide, a significant amount of energy remains on the surface, turns into heat and quickly flashes out into space. In order to find this missing amount of gravitational energy, we will reason the following way. Imagine a planet that has a mass equal to the Earth's mass, but consists of a non-condensable substance. Let's assume that all of the gravitational energy released during the formation of this planet was emitted into outer space. It has thus cooled and is cold. Its density is assumed to be equal to the average density of cooled rocks on the Earth's surface $\rho = 3 \, \text{g/cm}^3$. We will find the radius of this hypothetical planet:

$$R_0 = \left(\frac{3 M_E}{4 \pi \rho} \right)^{\frac{1}{3}} = 7800 \, \text{km} (4846.70 \, \text{mi}). \qquad (3.19)$$

In this case, its gravitational energy will be equal to:

$$E_0 = -\frac{3 G M_E^2}{10 R_0} = -9.2 \times 10^{31} \, \text{J}. \qquad (3.20)$$

Therefore, the thermal energy that remained in the Earth's interior when it formed will be equal to:

$$E_E = \left| E_g - E_0 \right| = 5.3 \times 10^{31} \, \text{J}. \qquad (3.21)$$

This energy was enough to melt the Earth's core and provide a flow of heat from its interior to its surface. Estimates suggest that within the first billion years of the Earth's existence about half of this energy was transferred from its depths to its surface.

Now we will focus in greater detail on the structure of the Earth's interior. We will start at the surface and move inward to the depths of our planet (Fig. 3.18).

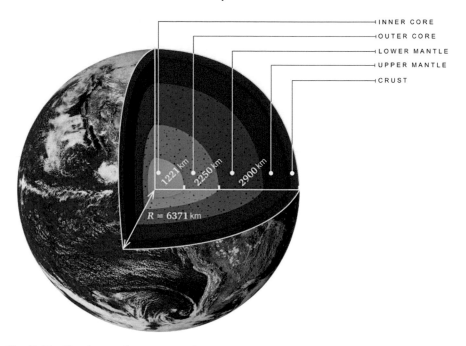

INNER CORE
OUTER CORE
LOWER MANTLE
UPPER MANTLE
CRUST

1221 km
2250 km
2900 km

R = 6371 km

Fig. 3.18 The internal structure of the Earth

The first layer—the Earth's outer shell—that we have to work our way through is the Earth's crust. Its thickness varies from several kilometers (miles) under the ocean to tens of kilometers (miles) in the mountainous regions of the continents. The Earth's crust is made up of stone and its chemical composition is primarily silicon oxides, aluminum, iron and alkali metals. The crust of the continents is covered with a layer of sedimentary rock, under which a layer of granite is found and below that there is a layer of basalt. The crust under the ocean under the sedimentary layer contains only one basaltic layer.

The next layer is mantle. It accounts for about 67% of the total mass of the Earth.

The temperature of mantle is thousands of degrees. Since there is a significant difference in the temperature of the outer and inner layers of mantle, the substance of which it is composed is constantly being mixed together. Hot masses of matter rise up, cool down and then sink back down. Boiling water in a tea kettle acts the same way, but in the Earth's interior circulation is much slower because the substance that makes up mantle is significantly thicker than water. The solid upper layer of mantle combined with the Earth's crust is called the *lithosphere*. The lithosphere is the rigid shell of the Earth.

At a depth of about 2900 km (1801.98 mi), the Earth's mantle borders its core. Studies have revealed that the Earth's core is divided into two areas: the so-called outer core, which is a fluid layer, and the solid or inner core. The transition between the liquid and solid regions is located approximately 5150 km (3200 mi) beneath the Earth's surface. The outer liquid core is made up of churning molten metal—iron and nickel—that flows up and down. Since these metals conduct electricity well, scientists associate the origin of the Earth's magnetic field with this giant generator. The inner core is solid because of tremendous pressure in the center of the Earth. Estimates suggest that that the amount of pressure in the Earth's core reaches three million atm and its mass density is about 12 g/cm^3. This pressure makes the substance of the inner core solid despite the very high temperature there.

3.5   Movement of the Earth's Crust

Mantle has a high viscosity; it can be compared to a thick resin. The reason for this is the high temperature and pressure found in this area. The lithosphere, which is the outer shell of our planet, "floats" in thick mantle and is slightly submerged in it under its own weight.

The fact that the density of basalt is higher than the density of granite results in an interesting phenomenon called *continental drift*. The Earth's continents actually move with respect to one another and what is more, the speed of this movement is several centimeters (inches) each year. It is believed that at one time there was a single continental mass called Pangaea, but sometime later it broke apart into separate continents. One of the first reconstructions of the location of the continents before continental drift is shown in Fig. 3.19.

Numerous facts support the hypothesis of continental drift, the most notable of which is that the edges of the continents all fit together well. South America and Africa match, as does the outer contour of North America with the contour of Greenland and northwest Europe. With a certain amount of simple rotation, we can join South America, Africa, Antarctica, the Southeast Asian Peninsula and Australia together in several ways. This is easy to prove. All one has to do is cut out the contours of the continents from a map and put Pangaea together like a puzzle (Fig. 3.20).

The patterns of the internal and geological structure of the continents in the eastern regions of North America and northwest Europe, as well as of South America, Africa and Antarctica are consistent. The conformity of the geological structures of South America and Africa (Fig. 3.21) confirms that

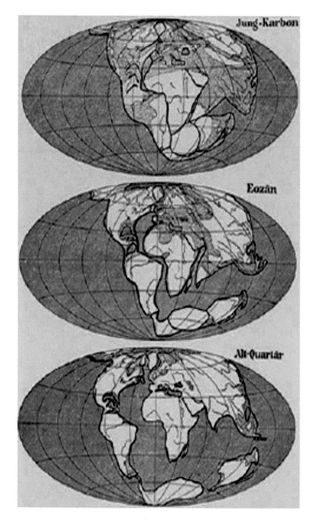

Fig. 3.19 The movement of the continents according to the opinion of Alfred Wegener. Illustrations from the book *The Origin of Continents and Oceans*, 1929

these two continents that are separated today by the Atlantic Ocean used to be one.

The similarity of types of flora and fauna in different continents is striking. For example, remains of animals have been found in Africa that bear a striking resemblance to ones in South America. Identical prehistoric types of plants were found in Europe and North America and the same primitive type of ferns were found in India and Antarctica. A 200-million-year-old fossilized reptile, which was found in rocks in the middle of the last century near the South Pole, turned out to be related to creatures native to South Africa.

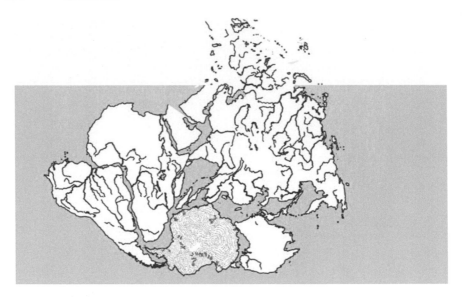

Fig. 3.20 A reconstruction of the possible form of an early supercontinent

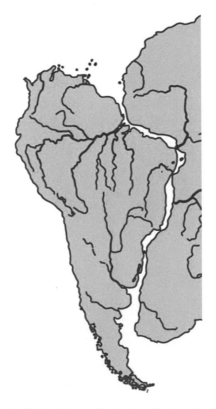

Fig. 3.21 The alignment of the shape of the South American and African coastline

Today scientists believe that the continents themselves do not move, but rather large areas of the Earth's crust or the so-called tectonic (or lithospheric) plates move, which include both continents and large sections of the ocean floor. There are only six main plates: Eurasian, North American, South American, Indo-Australian, Pacific and Antarctic and among them there are several smaller ones (Fig. 3.22).

The boundaries of the plates lie in the oceans, and earthquakes and volcanic eruptions occur when they slip past one another.

What is the source of energy that forces tectonic plates to move? The answer is that this source is primarily gravitational energy and the amount of it that is used is determined by the gravity of rocks that have different densities. It is the force of gravity that forces denser substances to drop down, while those that are less dense rise up as if floating to the surface. It is possible that the strength of lunar and solar tides that acts upon the Earth's crust also exerts a specific influence on it.

It is difficult to imagine floating continents because our life experience says that there is nothing more stationary than the mountains, fields and lakes that surround us. In order to see the changes caused by the movement of the continents, not only will we have to wait for several generations to pass, but also for new species to evolve.

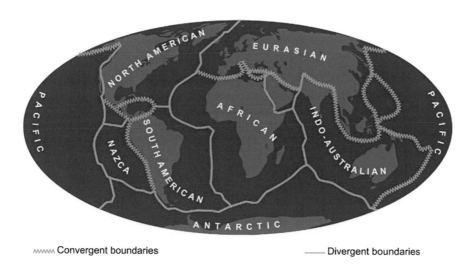

〰〰 Convergent boundaries   —— Divergent boundaries

Fig. 3.22 Tectonic plate boundaries

3.6   Volcanoes and Geysers

From our perspective, the Earth is a solid motionless surface. However, volcanic eruptions provide evidence that the matter inside of it is not solid at all and we live on large basalt-granite "rafts;" specifically, tectonic plates that freely float in the bottomless ocean of molten matter. Why is it that on the Earth's surface we can occasionally witness a powerful eruption of the matter inside of its inner core?

It is surprising, but the rationale for the eruption of volcanoes is the same as the reason why a bottle of sparkling water that has been shaken spews out liquid when opened. We all know that if a bottle of a carbonated beverage (e.g., soft drinks or kvas[2]) (Fig. 3.23a) is carefully opened, a mist appears and sometimes there is a loud pop and foam forms. This is dissolved gas escaping from a drink. But if you shake or heat a bottle before opening it, a strong spray of gas and liquid, which is very difficult to contain, shoots out of it (Fig. 3.23b). The fact of the matter is that when a bottle is opened, the pressure in the bottleneck decreases and the gas that dissolved in the liquid begins to be very actively released. This is called *degassing*. When gas bubbles rise to the surface, they propel some of the liquid out of the bottle.

Here is what happens inside of the volcano. The mantle that has been heated by the Earth's core rises up to the base of the lithosphere and then after having cooled down, once again sinks back down toward the center of the Earth. When this mantle is moving, it drags along pieces of lithospheric plates. Those sections of plates that sunk down below the others end up in

Fig. 3.23 Degassing while opening: **a** a can of a carbonated drink, **b** a bottle of sparkling water that has been shaken

---

[2] A traditional fermented drink made from rye bread, which is popular in Russia and other eastern European countries (translator's note).

a high temperature area and begin to melt. This is how magma is formed—it is a thick mixture of molten rocks combined with gases and water vapor. Magna has a temperature of about 800 °C (1472 °F).

Magma has the consistency of thick porridge and means *dough* in Greek.

Magma accumulates in empty areas below the Earth's surface in magma chambers (Fig. 3.24). Consequently, because of high pressure these areas fill up with "highly carbonated" molten silicone oxide.

Imagine that at a spot not far from the magma chamber, the Earth's crust happened to be loosely plugged up. Magma pushes out the volcanic "plug" and erupts out of the bowels of the Earth. What can aid magma in shooting to the surface? The answer is that geological phenomena such as a small earthquake, the melting of glacier caps that cover the vent of a volcano, a landslide and even very heavy rains can do this.

When a volcano "loses its roof," the pressure in the upper part of the magma chamber rapidly drops. Pressure remains high in the lower part of the magma chamber where high amounts of dissolved gases are still part of the magma. Gas bubbles begin to be intensively emitted from the volcanic vent. The higher the area in the throat of the volcano, the lower the pressure and the more gas bubbles there are. Large and small bubbles quickly rise up and carry away the thick magma matter. A large foamy mass— pumice,

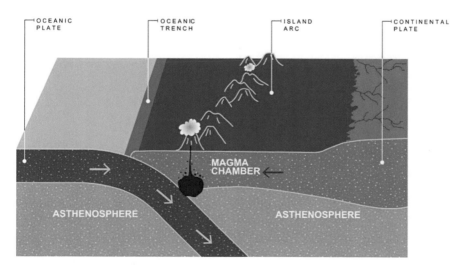

Fig. 3.24 The emergence of a magma chamber in a depression zone of a crustal block (an oceanic plate) under another chamber (a continental plate)

which everyone knows so well—forms on the Earth's surface at the vent of the volcano. Because of the large number of pores in which the dissolved gases are located, this volcanic stone foam is even lighter than water! Magma degassing culminates on the surface. While rising along the throat of the volcano, magma turns into lava, which is gas-depleted magma, as well as hot gases, water vapor, ash and rock fragments. All of this together with fire and smoke spew onto the Earth's surface and loud explosions are also heard. Although a stunning sight, volcanic eruptions are also dangerous (Fig. 3.25).

Why do explosions often occur during volcanic eruptions? The reason is that the gases that are released during a volcanic eruption (i.e., hydrogen, methane, carbon dioxide and sulfur) are combustible, which means that at high temperatures they ignite and explode in the throat of the volcano. The force of a volcanic explosion can be so strong that after the eruption a huge cauldron-like hollow or *caldera* is all that remains where a mountain had

Fig. 3.25 A gas explosion coupled with the release of volcanic ash during a volcanic eruption

Fig. 3.26 The caldera caused by Gorely Volcano in Kamchatka (Russia)

stood (Fig. 3.26). If the eruption continues, a new volcano begins to grow right in the caldera.

Thus, the degassing of magma is the physical reason for a volcanic eruption.

Now we understand why zones of volcanic activity are located near tectonic plate boundaries and on the shores of seas and oceans (Fig. 3.27) where there are faults in the Earth's crust because it is in those places that there are anomalous weak spots in the Earth's solid outer layer.

But what happens to a volcano when degassing culminates, gas pressure in the magma chamber decreases and the volcano has stopped erupting? If the magma chamber has not filled up with a new amount of magma, then the volcano becomes dormant. But it can wake up if the movement of geological plates continues and the magma chamber fills up again with magma. There have been cases of volcanoes erupting after they had been dormant for hundreds of years. Volcanoes that erupted at least once and can wake up are called *dormant*, while *extinct volcanoes* are those that erupted millions of years ago and are not expected to ever erupt again.

After a volcano has erupted it begins to cool down and its throat fills up with hardened lava. But if the temperature in the magma chamber remains high enough, volcanic gases (i.e., steam vents) or columns of hot water (i.e., geysers) will continue to erupt onto the surface. This type of volcano is considered active because at any time a large amount of magma can accumulate in the magma chamber again and then another eruption will occur. A volcano becomes completely extinct only when the tectonic plates in that area stop moving altogether.

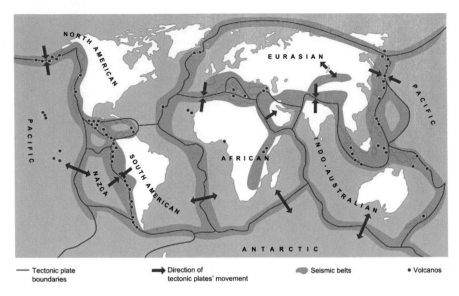

Fig. 3.27   The location of volcanic zones on the Earth's surface

Fig. 3.28   The extinct volcano "Arthur's Seat" in Edinburgh (Scotland)

Edinburgh, the capital of Scotland, stands on an ancient volcano that erupted more than 300 million years ago (Fig. 3.28).

When a volcano erupts, the greatest danger that people living near it face comes from the emission of lava, ash and volcanic gas. But even if people are a great distance away from a volcano, they are not out of harm's way because they may be hit by a so-called *volcanic bomb*. This is a mass of lava and stones that weighs up to several tons (Fig. 3.29) and can fly up to 25–30 km (15.53–18.64 mi).

Fig. 3.29 A volcanic bomb near a volcano on the island of Lanzarote (Canary Islands, Spain)

The Range of a Stone Spewed from a Volcanic Vent Let's say that a large stone flies out of a volcano vent. The volcano is located at the height H above ground at the angle β to the horizon with the original velocity v_0. When we formulate an equation of the stone's motion into the field of gravity and disregard drag, it is easy to set up an equation to find this range of motion. The trajectory has a parabola:

$$\frac{g}{2v_0^2(\cos\beta)^2}x^2 - \text{tg}\,\beta x - H = 0. \tag{3.22}$$

In this situation the distance from the volcano to the place where the stone falls will be determined by the formula:

$$L = \frac{v_0^2(\cos\beta)^2}{g}\left(\text{tg}\,\beta + \sqrt{(\text{tg}\,\beta)^2 + \frac{2gH}{v_0^2(\cos\beta)^2}}\,\right). \tag{3.23}$$

The maximum distance that is obtained is $\beta = 45°$. On the assumption that $v_0 = 300$ m (984.25 ft)/s and $H = 1$ km (0.62 mi), we get $L = 10$ km (6.21 mi).

The energy from a volcanic eruption is tremendous. As a case in point, the energy from one of the largest eruptions—Krakatoa Volcano in Indonesia in 1883—measured about 10^{20} J. The world's most powerful hydroelectric power plant, Three Gorges Dam in China on the Yangtze River, delivers 22 GW of power. It is easy to figure out that in order for it to produce an amount of energy comparable to that of a volcanic eruption, this hydroelectric power plant would have to work for almost two years! Considering the fact that eruptions usually last only a few days, one can understand that given volcanoes' colossal power, they are one of the most impressive and dangerous types of natural phenomena that exist.

The Energy from a Volcanic Eruption The energy from a volcanic eruption can be estimated as follows. The bulk of this energy is thermal energy, which is dependent upon the heating and melting of eruptive rocks. It is equal to:

$$E = V_{lava}\rho_{lava}(c_{lava}\Delta T + \lambda_{lava}).$$

(3.24)

We need to know the volume of the lava that flowed out during the eruption and the difference between its temperature and the air temperature. The thermodynamic parameters of the lava are known: density $\rho_{lava} = 2700\,kg/m^3$, heat capacity $c_{lava} = 840\,\frac{J}{kg\,K}$ and heat of fusion $\lambda_{lava} = 3.5 \times 10^5$ J/kg. If the volume of the erupted substance is 1 km^3 and the lava temperature is 1200 °C (2192 °F), then the energy of eruption turns out to be about $E \approx 4 \times 10^{18}$ J.

Types of Volcanic Eruptions There are six known types of volcanic eruptions (Fig. 3.30). They may alternate during the lifespan of the same volcano and even during one of its stages of volcanic activity.

A Hawaiian eruption (Fig. 3.30a) is characterized by a constant release of magma. Lava from the volcanic vent erupts in a fountain and spreads over the mountain's slopes.

A Strombolian eruption (Fig. 3.30b) has repeated explosions, frequent eruption of volcanic bombs, slag, pumice and ash. At the same time, a small amount of flowing lava is produced.

In a Peléan eruption (Fig. 3.30c), gas escapes from the crater with explosive force. The lava that has accumulated and cooled around the crater forms a plug. After lava is released, incandescent clouds appear.

A Plinian eruption (Fig. 3.30d) is characterized by explosions that create a giant umbrella-shaped cloud of ash above the crater, which rises several kilometers (miles) into the sky.

A Vulcanian eruption (Fig. 3.30e) is characterized by frequent rhythmic explosions of volcanic gases.

Powerful gas emissions occur through the vent that is clogged with debris from previous eruptions.

A Surtseyan eruption (Fig. 3.30f) is an underwater eruption. A layer of compressed ash, known as tuff, forms around the crater. Tuff gradually rises up from the sea and forms a volcanic island.

Fig. 3.30 Known types of volcanic eruptions

There are also less dangerous demonstrations of the Earth's activity. We have already mentioned that the magma's high temperature heats groundwater aquifers. This, in turn, causes hot springs to reach the Earth's surface where steam and water are ejected. This kind of hot spring is called a *geyser*.

Each geyser resembles a tube, that is, a channel that runs deep into the crust under the Earth's surface and due to an uneven heat distribution in the ground, the temperature inside of the tube increases with depth. The tube can be up to 3 m (9.84 ft) in diameter and have a length of more than 25–30 m (82–98.43 ft). Immediately following the geyser's eruption water runs out of the tube and leaves it empty. After a certain period of time the tube once again fills up with water, which begins to boil and an eruption occurs; specifically, a fountain of boiling water shoots up. In order to comprehend this process, it is important to understand that the boiling point of water depends on pressure (Fig. 3.31). If the temperature of the boiling water is 100 °C (212 °F) at a pressure rating of 1 atm, then when pressure is increased, for example to 2 atm, water's boiling point increases to 120 °C (248 °F). However, when pressure is decreased, the boiling point decreases. Mountain climbers who have to boil tea high in the mountains understand this concept very well.

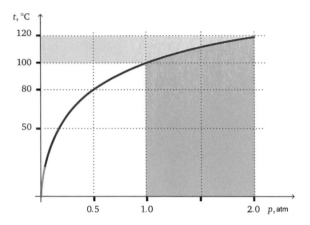

Fig. 3.31 Dependence of water's boiling point on pressure

For the purposes of illustration, let's consider a geyser with the following parameters. The height of the tube is 20 m (65.62 ft); the water temperature at the upper edge of the tube is 85 °C (185 °F), reaching 122 °C (252 °F) and 126 °C (259 °F) at a depth of 13 m (42.65 ft) and at the base of the tube, i.e., at a depth of 20 m (65.62 ft), respectively (Fig. 3.32a). As we go deeper, the hydrostatic pressure of the column of fluid, which is equal to $\rho_w g H$, increases. Nevertheless, the water temperature is always slightly lower than the boiling point at this amount of pressure. Now let's imagine that steam has entered into one of the side channels, which is located at the height $H = 13\,\text{m}\,(42.65\,\text{ft})$. This vapor has pushed the water to a certain height, let's say to $\Delta H = 2\,\text{m}\,(6.56\,\text{ft})$. It is important to note that part of the water from the top half of the tube will overflow into the basin outside of the geyser (Fig. 3.32b). At the same time, the water that was at a depth of 13 m (42.65 ft) when the temperature was 122 °C (252 °F), will be at an 11 m (36 ft) depth where the boiling point is 121 °C (250 °F). This water will instantly start to boil and when it does, vapor bubbles will form that will push the water out of the geyser's tube with even greater force. The less water that remains in the tube, the lower the hydrostatic pressure, and, hence, the boiling point will be.

The pressure will eventually decrease so much that all of the water in the tube will begin to boil (Fig. 3.32c). When this happens, a large amount of steam will immediately appear, which will quickly move up the tube and in the process push out any remaining water, as well as all of the water from the basin.

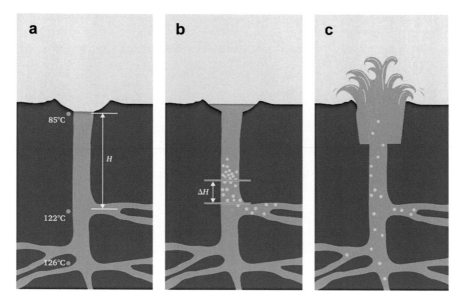

Fig. 3.32 An illustration of the formation of a geyser

After all of the steam has escaped from the tube, cold water will flow back in. The next eruption will occur only after the water in the tube has once again heated up to a temperature close to the boiling point.

An example is the Valley of Geysers in Kamchatka (Fig. 3.33). This is a most unique place where more than two dozen large geysers and several hundred small ones are located on an area measuring 6 km^2.

3.7 Earthquakes

We have already said that tectonic plates can be imagined as stone rafts floating on molten mantle on the Earth's surface. When these plates move, their edges can, of course, collide. They then get deformed at the points of collision—one slides under the other and they become bent (Fig. 3.34). When this deformation reaches a certain critical level, slippage occurs at the plate boundaries and oceanic plates forcefully go back to their undeformed state. As a result of this movement, tremendous forces of deformation—known as *tectonic forces*—are activated, which give rise to major geological changes such as the formation of mountains and the occurrence of earthquakes.

Earthquakes are vibrations of the Earth's surface caused by tectonic processes, that is, the natural movement of tectonic plates.

Fig. 3.33 The Valley of Geysers (Kronotsky Nature Reserve, Kamchatka, Russia)

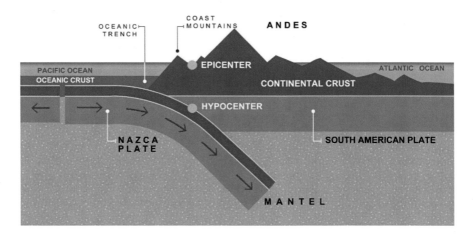

Fig. 3.34 An illustration of the earthquake in Chile on May 22, 1960

The location in which tectonic stress is instantly released is called the *hypocenter* and the location directly above it on the Earth's surface is called the *epicenter*.

Part of the released potential energy that comes from deformation heats rocks and about 10% of it is converted into seismic waves. These waves run along plates in different directions and when they reach the Earth's surface, they cause rocks to crack and shift (Fig. 3.35).

There are several types of seismic waves (Fig. 3.36).

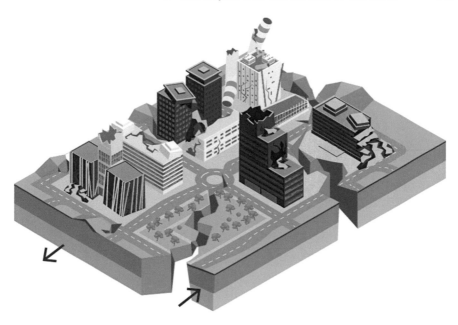

Fig. 3.35 Cracks that form in the Earth's surface during earthquakes

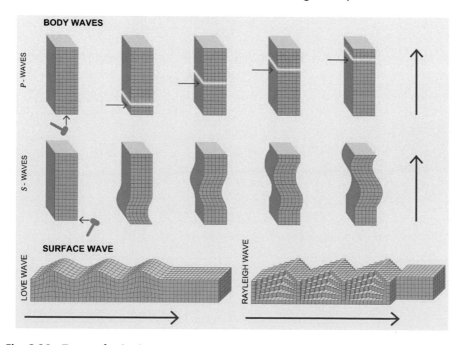

Fig. 3.36 Types of seismic waves

In an earthquake's hypocenter body waves are generated that propagate in all directions. When a body wave reaches the Earth's surface, it stimulates a surface seismic wave.

A longitudinal body wave is called a *P wave*. These waves travel at a speed of 7–11 km (4.35–6.83 mi)/s and act similar to a shock wave during an explosion, that is, they are followed by a loud boom.

A transverse body wave is called an *S wave*. These waves propagate at a speed of 3–8 km (1.86–4.97 mi)/s. They displace the surface of the soil by shaking it horizontally and vertically and cause significant damage.

Thus, P waves, which propagate faster than S waves, reach an observer earlier. If we measure the time between the arrival of these waves Δt, then since we know that their speed is v_P and v_S, we can easily estimate the depth at which the hypocenter of this particular earthquake is located:

$$H = \frac{v_P v_S}{v_P - v_S} \Delta t. \tag{3.25}$$

When a body wave reaches the Earth's surface, it transfers part of its energy in order for surface seismic waves to form, which move along the Earth's surface at a speed of 1.5–2 km (0.93–1.24 mi)/s. The amplitude of body waves decreases in inverse proportion to the distance to the epicenter, while the amplitude of surface waves is slower and is inversely proportional to the square root from this distance. Thus, this is exactly why surface waves cause the greatest destruction at places on the Earth's surface that are far from the epicenter.

Surfaces waves are divided into Love and Rayleigh waves (see Fig. 3.36). The former are a superimposition of transverse waves and their particles vibrate horizontally. This vibration is perpendicular to the direction of wave propagation. The latter are a superposition of longitudinal and transverse waves. Their particles move along ellipses in a plane in which the velocity vector and the normal to the Earth's surface lie. The amplitude of Rayleigh and Love waves decreases in inverse proportion to the square root of distance to the source of the vibration.

The amount of energy released during an earthquake is described as magnitude. By convention the magnitude of an earthquake M is determined by using the following formula:

$$M = \frac{2}{3} \lg E - 3, \tag{3.26}$$

where lg is the common logarithm and E is the seismic energy measured in joules.

The magnitude of earthquakes is measured on a scale from 0 to 9.5. The minimum value $M = 0$ corresponds to $E = \sqrt{10^9}\,\text{J} \approx 3 \times 10^5\,\text{J}$ of tectonic energy. This is approximately the amount of energy that is released when a boulder weighing 500 kg (1102 lbs) falls from a height of 6 m (19.685 ft). Clearly this fact is seismically insignificant. The maximum value of magnitude $M = 9.5$ corresponds to $E = 10^{19}\,\text{J}$ of energy. This is the same amount of energy that would be released, if, for example, the Earth were to collide with a large asteroid.

If two earthquakes occur with a level of magnitude that differs by one, $M_2 - M_1 = 1$, then their tectonic energy will differ in relation to the formula $\lg\left(\frac{E_2}{E_1}\right) = \frac{3}{2}$ or $\frac{E_2}{E_1} = 31.6$. Hence, increasing the magnitude of an earthquake by 1 corresponds to a 30-fold increase in its tectonic energy.

It is possible to estimate the destructive ability of an earthquake on the Earth's surface by using an international 12-point scale. According to this scale, an earthquake is considered weak if it is within a I–IV-point span, strong if it is within a V to VII-point span and destructive if it is higher than VIII points.

A scale of earthquake intensities with corresponding magnitudes is provided in Table 3.2.

People often confuse magnitude and intensity. Although they are related measures, they are very different.

The connection between the magnitude and intensity of an earthquake's epicenter is determined by the depth at which its hypocenter is located. A strong earthquake that strikes deep underwater may do less damage to the Earth's surface than a weak one near its surface.

> When newscasters speak about earthquakes, they often make a mistake when they say that there was an earthquake with a magnitude of five points on the Richter scale. Magnitude is, however, a non-dimensional value and is not measured in points. Therefore, it is correct to say, "A magnitude 5.0 earthquake."

Weak earthquakes happen all of the time. But, thankfully, strong ones—disastrous natural phenomena that sometimes cause tremendous destruction and take human life—are very rare. The deadliest earthquake struck China in 1556 and claimed the lives of more than 800,000 people. The most powerful earthquake, which was a magnitude 9.5, hit Chile in 1960 (Fig. 3.37). It is necessary to study earthquakes in order to learn how to predict them and take timely measures to keep people safe.

Table 3.2 A scale of earthquake intensities

Magnitude	Strength of earthquake	Brief description
1	Not felt	Recorded only by seismometers
2	Just perceptible	Recorded by seismometers. Felt only by those who are in a state of rest on the upper floors of buildings, as well as by very sensitive pets
3	Slight	Felt only inside of some buildings, similar to the vibration caused by a passing truck
4	Perceptible	Characterized by a slight rattling sound, shaking of dishes and windowpanes and creaking of doors and walls. These tremors are felt by most people who are inside of buildings
5	Rather strong	Felt by many people outside, but everyone inside. Buildings shake and furniture moves. Clock pendulums stop. Glass panes and plaster crack. Those who are asleep wake up. Felt by people outdoors. Thin tree branches swing back and forth. Doors slam shut
6	Strong	Felt by everyone. Many people run outside in fear. Pictures fall off walls. Pieces of plaster break off
7	Very strong	Damage (i.e., cracks) occurs in the walls of stone houses. Earthquake-resistant structures, as well as wooden and wattle constructions are not damaged
8	Destructive	Cracks on steep slopes and moist soil form. Statues may shift or tip over. Houses are severely damaged
9	Ruinous	Stone houses are severely damaged and destroyed. Old wooden houses tilt
10	Disastrous	Cracks in the soil, sometimes up to one meter (3.28 ft) wide appear. Slopes give way to landslides and avalanches. Railroad tracks are twisted
11	Catastrophic	Wide cracks in the top layers of soil form. Numerous landslides and avalanches occur. Stone houses are almost completely destroyed. Railroad tracks are severely twisted and buckle

(continued)

Table 3.2 (continued)

Magnitude	Strength of earthquake	Brief description
12	Severely catastrophic	Topographic changes of monumental proportions take place. Numerous cracks, avalanches and landslides occur. Waterfalls and weirs on lakes appear, rivers change their course. Not a single structure can withstand this type of earthquake

Fig. 3.37 The aftermath of the most powerful earthquake in the history of mankind, which struck the city of Valdivia (Chile) on May 22, 1960: **a** local residents' destroyed homes, **b** aerial photography of the coastline of Chiloé Island, which was completely destroyed by a tsunami that followed the earthquake

Further Reading

1 Ammon, C.J., Velasco, A.A., Lay T., Wallace, T.C.: Foundations of Modern Global Seismology, 2nd edn. Academic Press, Cambridge (2020)
2 Byalko, A.V.: Our Planet the Earth. MIR Publisher, Moscow (1983)
3 Gregory, T.: Meteorite: How Stones from Outer Space Made Our World. Basic Books, New York (2020)
4 Kump, L., Kasting, J., Crane, R.: Earth System, 3rd edn. Pearson, London (2019)

5 Lowrie, W., Fichtner, A.: Fundamentals of Geophysics, 3rd edn. Cambridge University Press, Cambridge (2020)

6 Lutgens, F.K., Tarbuck, E.J., Tasa, D.G.: Essentials of Geology, 12th edn. Pearson, London (2014)

7 Maral, G., Bousquet, M.: Satellite Communications Systems: Systems, Techniques and Technology, 4th edn. Wiley, Hoboken (2002)

8 Ohnaka, M.: The Physics of Rock Failure and Earthquakes. Cambridge University Press, Cambridge (2018)

9 Palmer, D.F., Allison, I.S.: Geology: The Science of a Changing Earth. McGraw-Hill Companies, New York (1980)

10 Rubtsov, V.: The Tunguska Mystery. Springer, Berlin (2009)

11 Wegener, A.: The Origin of Continents and Oceans. Dover Earth Science (2018)

4

The Earth's Atmosphere

Abstract The fourth chapter is devoted to the physics of the Earth's atmosphere and the major atmospheric phenomena. We consider the structure and chemical composition of the Earth's atmosphere, the color of the sky, and such physical phenomena as rainbows and mirages. Then we discuss the movement of gases in the Earth's atmosphere and the reasons for winds on the Earth. We estimate the average speed of the wind and the main features of global and local winds on the Earth.

What exactly is a planet's atmosphere? It is its gaseous envelope held by gravity. The atmosphere of different planets in the Solar System varies greatly just as the planets themselves do too. The atmosphere of Mars and Venus consists mainly of carbon dioxide, while the atmosphere of Jupiter and Saturn are made up primarily of helium, hydrogen, methane and ammonia. There is practically no atmosphere on Mercury.

The Earth's atmosphere is largely composed of nitrogen (about 78%) and oxygen (about 21%). The most important elements needed for life to exist are water vapor and carbon dioxide, which must be present in the atmosphere, although together their percentage falls short of 0.01% of the total mass of the gaseous envelope. The mass of the entire atmosphere is 5.3×10^{18} kg, which is a million times less than the Earth's mass.

© The Author(s), under exclusive license to Springer Nature
Switzerland AG 2023
D. Livanov, *The Physics of Planet Earth and Its Natural Wonders*,
https://doi.org/10.1007/978-3-031-33426-9_4

The Atmosphere's Mass We know that the atmospheric pressure at the Earth's surface is $p_0 = 10^5$ Pa. The surface area of our planet is $S = 4\pi R_0^2$, where $R_0 = 6370$ km (3958.13 mi), which is the average radius of the Earth. The total amount of force that acts on the Earth's surface is $F = 4\pi R_0^2 p_0$; however, according to Newton's second law, this is the mass of the Earth's atmosphere m_a multiplied by the acceleration of gravity g. Although the value of g decreases with height, it is possible to derive its value at the Earth's surface in order to get an estimate. Thus, we get:

$$m_a = \frac{4\pi R_0^2 p_0}{g} = 5.3 \times 10^{18} \text{ kg}. \tag{4.1}$$

Although we do not notice it, we live at the bottom of a deep ocean of air. Despite the apparent "windiness" and ephemerality of this ocean, it firmly protects us from the deadly effects of outer space: different types of high-level radiation, frigid cold air and streams of small particles burning in the atmosphere. The bottom line is that the atmosphere produces the pressure that is essential for us to live and gives us oxygen, which produces the energy that all living organisms need to survive. If we did not have an atmosphere, our landscapes would bear a striking resemblance to those of the Moon—they would be lifeless, depressing and dotted with meteorite craters. If we did not have an atmosphere, we would not hear any sounds, see the blue sky, a colorful rainbow or delight in watching birds fly.

It is very important to understand what is happening in the atmosphere and to be able to predict what will take place there because almost all human activity depends on the weather, which is determined by the atmosphere.

4.1   The Structure of the Atmosphere

At first glance it seems that the atmosphere is structured very simply—it is composed of air. Moreover, air pressure is greater closer to the ground, but the higher we go from the surface of Earth, the lower the air pressure. The atmosphere ends in the exact spot where there is practically no pressure at all. Water vapor, which forms clouds and fog, is in the air very close to the Earth itself. In actuality, the structure of our atmosphere is much more complicated and fascinating (Fig. 4.1).

The troposphere is the lowest layer of the atmosphere. Its thickness is between 8 and 18 km (5–9 mi).

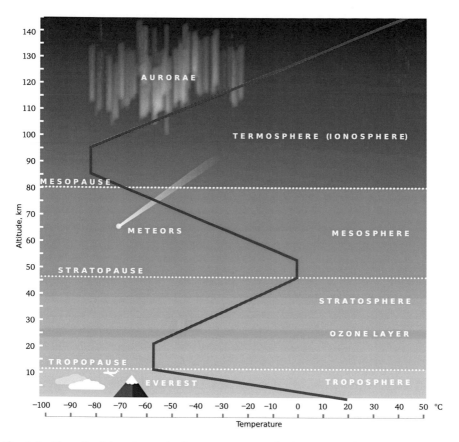

Fig. 4.1 The Earth's atmospheric structure and the atmospheric temperature depending on altitude (red line)

This is, in fact, where we live, and even when flying in an airplane, we stay within the limits of the troposphere. It is this layer that is responsible for our weather: almost all water vapor is found in the troposphere, i.e., rain, storms, hurricanes and clouds form here. It is specifically the temperature of the troposphere that is mentioned in weather reports; on the Earth's surface, it ranges from about − 70 to + 50 °C (− 94 to 122 °F). Eighty percent of the entire mass of the atmosphere is found in the troposphere.

> The stratosphere is located above the troposphere at an altitude of 8–50 km (4.97–31.06 mi).

The remaining 20% of atmospheric mass is concentrated in the stratosphere. It can be reached on a high-altitude balloon, which is a massive balloon filled with helium. Man's difficult journey into space began by studying the stratosphere. People who dared to fly that high were called stratosphere pilots, and in the 1930s they were rightly considered heroes. They were the first ones to go above the troposphere where it is impossible to breathe without an oxygen mask, where the air is so thin that it is necessary to wear a spacesuit, where the sky is a violet color and the wind constantly blows at a speed of 300 km (186.41 mi)/h.

The ozone layer, which keeps a large portion of ultraviolet radiation from passing through to the Earth's surface, is located in the stratosphere. The ozone layer is about 5 km (3.11 mi) thick, but there is relatively little ozone in it. In fact, if all ozone atoms were compressed to the pressure of the air at sea level, this layer would be only a few millimeters (inches) thick! Nevertheless, the ozone layer is very important because it protects all living organisms on our planet.

> The mesosphere is located at an altitude of 50–80 km (31.07–49.71 mi) from the Earth.

From the point of view of any living organism, this environment is not much different from outer space: the air density here is low, in fact it is 200 times lower than near the Earth's surface. Stars and a dazzling white Sun shine in the black sky. The air temperature drops to − 75 °C (− 103 °F) and lower here. It is impossible to survive in such an empty place. However, for celestial bodies that make their way here from outer space, it is not a completely barren area because most meteors burn up in the mesosphere.

> The thermosphere begins above the mesosphere, and it is the longest part of the atmosphere. It starts at an altitude of 90 km (55.92 mi) and extends to an altitude of more than 800 km (497.10 mi) from the Earth.

The air in the thermosphere is obviously even thinner than in the mesosphere. This part of the atmosphere is interesting because the temperature here is unexpectedly high: from 200 to 2000 K depending on solar activity. It heats up due to ultraviolet radiation from the Sun, but do not be mistaken about the high temperature here—it is impossible to get sunburned. This is due to the fact that when the air is so rarefied, the whole concept of temperature is very arbitrary. We are mistaken when we consider the statistical meaning of temperature as a measure of average kinetic energy. This is because individual particles that move within a rarefied space reach a very high speed. However, it is incorrect to speak about the temperature of individual particles because when they interact with a substance (e.g., the outer shell of a spacecraft), they cannot heat it up.

The thermosphere is the layer where orbital space stations, spaceships and satellites orbit the Earth. Why is the thermosphere the preferred place for space travel? The reason is that there are too many air molecules below the thermosphere, and they significantly slow down the movement of space stations. Cosmic rays, on the other hand, which are extremely harmful to people, are located above the thermosphere.

> The exosphere forms a conventional border between the atmosphere and outer space. Its starts more than 1000 km (621.37 mi) from the Earth.

The same gas particles that travel at a high speed are found here. For all practical purposes, they have left the atmosphere and are moving into outer space. They will never return to Earth.

This raises the question: Will all of the gas molecules of the Earth's atmosphere sooner or later fly away and cause the atmosphere to disappear? If this were to happen, the Earth would have the same fate as Mars, which lost almost all of its entire atmosphere. Now the amount of pressure on Mars is less than 1% of the amount on the Earth. Fortunately, the Earth's atmosphere will not suffer the same fate as Mars' atmosphere did!

Why Atmospheric Molecules Do Not Fly Away from the Earth The second cosmic velocity of the Earth is $\sqrt{2gR_0} = 11.2$ km (6.96 mi)/s. It is well known that the average thermal velocity of a gas molecule at the temperature T is $v = \sqrt{k_B \frac{T}{m}}$. We will calculate the thermal velocities for gas molecules, which make up the Earth's atmosphere, at the temperature 300 K, for example. This velocity is equal to 1.1 km (0.68 mi)/s for hydrogen, 0.8 km (0.50 mi)/s for helium and 0.3 km (0.19 mi)/s for nitrogen and oxygen. Indeed, gases escape into outer space because the velocity of certain molecules is different and within this group there is a small percentage of molecules that have a higher concentration of the first cosmic velocity. But atmospheric escape is happening very slowly.

It is also clear that the less the mass of a gas molecule, the easier it is to separate it from the Earth. This means that molecules of hydrogen and helium escape most actively into space. In actuality, our atmosphere loses 1 kg (2.20 lbs) of hydrogen every second. But hydrogen and helium do not only disappear, they are also produced on the Earth. Helium is formed due to the decomposition reaction of heavy elements—uranium, thorium, etc.—in the Earth's core, which provides about 5×10^6 kg of helium a year. Hydrogen enters the atmosphere when water molecules in the upper atmosphere break down because of solar radiation acting on oxygen and hydrogen. Given the fact that there is 1 kg (2.20 lbs) of hydrogen in 9 kg (19.84 lbs) of water, it stands to reason that there is more than enough water on the Earth to replenish the atmosphere with hydrogen. Moreover, it is also replenished by gases that are released from the hydrosphere and the lithosphere. Therefore, we don't have to worry about the fate of the Earth's atmosphere in the immediate future.

The atmosphere plays an important role in maintaining the thermal balance of the Earth. It is vital for us that it stays at zero; in other words, the amount of energy that the Earth absorbs in the form of radiation is the same amount that it should radiate. The reason for this is that if this balance is even slightly not maintained in one direction or another, it will end in disaster because everything will either burn up or freeze.

Let's take a look and see how the energy from the Sun is redistributed in the Earth's atmosphere. We will assume that the total energy of solar radiation is at 100%. Experiments show that far ultraviolet light (i.e., 1% of energy) is absorbed by the molecules of the exosphere and thermosphere, while near ultraviolet light (i.e., 3% of energy) is absorbed by ozone in the stratosphere. Additionally, some of the infrared spectral region (i.e., 4% of energy) is absorbed by water vapor in the upper layer of the troposphere. Thus, only 92% of solar energy passes through the Earth's atmosphere.

Almost half of this energy (45%)—mainly the blue visible portion of the spectrum—gets scattered evenly throughout the atmosphere primarily due to clouds, which give the sky its blue color. The remaining 47% of radiation reaches the Earth's surface as direct sunlight; of this amount, 7% is reflected back, while the remaining 40% heats the oceans and the Earth's surface.

Keeping in mind the amount of solar radiation needed to warm the atmosphere, of the total amount that reaches the Earth's surface, 65% of the energy that comes from it goes to heat our planet. This energy is radiated back by the Earth in the form of thermal (infrared) radiation and its emission power is 1.14×10^{17} W. The temperature that a completely black body with this type of radiation flow—257 K = $-16\,°C\,(3\,°F)$—would have is called the *radiation temperature* of the Earth. How is this consistent with the fact that the average temperature of the Earth's surface is much higher than this at about 15 °C (59 °F). The answer lies in the Earth's atmospheric structure. Despite the small amount of carbon dioxide and water vapor found in the atmosphere, carbon dioxide and water vapor molecules intensively absorb the infrared spectral region of electromagnetic radiation, which is what determines the low level of atmospheric transmission in this spectrum.

The phenomenon of heating the Earth's surface and the layer of the atmosphere adjacent to it is called the *greenhouse effect*.

A greenhouse (hothouse) works according to the following principle. The Sun's rays transfer heat inside of the greenhouse through a transparent covering, which causes the soil and air inside of the greenhouse to become warmer. But since glass is a good heat insulator, i.e., it has a low level of transparency for infrared radiation, heat cannot escape quickly. This means that the average temperature inside of the greenhouse stays higher for a longer period of time than the outside temperature does. Thus, the physical reason why the greenhouse effect is present in the Earth's atmosphere is that there is a high level of atmospheric transmission for visible solar radiation and a low level of atmospheric transmission for radiation in the infrared spectrum.

4.2   The Color of the Sky

In children's pictures, the sky is often drawn as light blue and the sun is yellow. This is how we perceive the world without giving any thought to why the sky is light blue, and not green, for example, on a clear day. There is an explanation in physics for the sky's color.

The sky is blue due to the fact that atmospheric gas scatters light with short wavelengths better than light with long wavelengths. According to Rayleigh's law, the intensity of light scattering I by charged particles, which are much smaller than the wavelength of radiation, are proportional to the fourth power of the radiation frequency, $I \sim \omega^4$.

This is the same as the scattering strength, which is inversely proportional to the fourth power of the wavelength of light $I \sim \frac{1}{\lambda^4}$.

Now we will turn our attention to the solar spectrum (Figs. 4.2 and 4.3).

The wavelength ratio of the red segment (650 nm) to the violet segment (440 nm) of the spectrum is 1.48. Although it is not a very large difference, if we raise it to the fourth power, we get a factor of 4.7. The fact is that violet light is scattered over four times more intensively than red. This is why our eyes see the Earth's atmosphere as blue. Why then isn't the sky violet? For two reasons. First, the intensity of the violet rays in the solar spectrum is much less than that of the blue rays (see Fig. 4.3). Second, the human eye perceives color emissivity differently (Fig. 4.4). In actuality, our eyes are much more sensitive to dark and light blue light than to violet light.

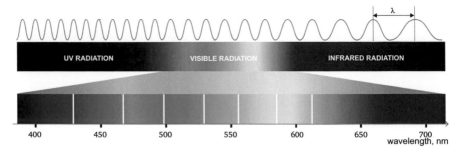

Fig. 4.2 Wavelengths of electromagnetic radiation in ultraviolet, visible and infrared ranges

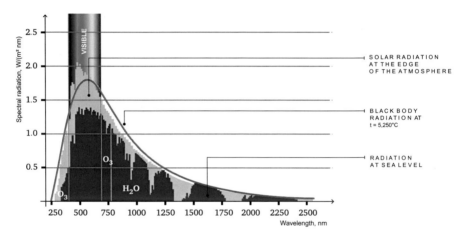

Fig. 4.3 The Sun's radiation spectrum at the edge of the Earth's atmosphere (in yellow) and at sea level (in red)

Why is that at sunrise and sunset, the Sun, the clouds it illuminates and the sky close to the horizon are red, orange and yellow? When it is low above the horizon, light travels in the atmosphere along a longer path than in the afternoon when the Sun is high above the horizon because sunbeams pass along a tangent to the Earth's surface. Consequently, yellow and red colors remain in the spectrum of light that passes through the atmosphere, but a great deal of blue and green light is scattered. Thanks to this, we get to see stunningly beautiful sunsets (Fig. 4.5).

About the Scattering of Light The concept of light scattering can be explained as follows. Light is an electromagnetic wave in which electrical density and magnetic strength depend on time in relation to simple harmonic motion $E, B \sim \sin \omega t$. The light particles that

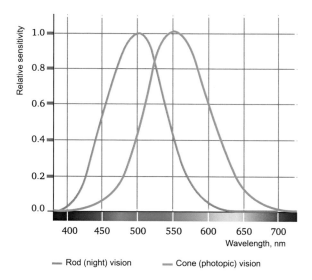

Fig. 4.4 The sensitivity of retinal cells in the human eye (rods and cones) to different parts of the emission spectrum

Fig. 4.5 A sunset on the Earth

get scattered are air molecules made up of nuclei and electrons. Since electrons are charged particles, in an electromagnetic wave with the electrical intensity E they begin to move according to the same principle of simple harmonic motion: $x(t) = A \sin \omega t$. The accelerated movement of this type of electron will be equal to $a(t) = -A\omega^2 \sin \omega t$. But a charged particle that moves at an accelerated rate becomes a source of electromagnetic radiation. Moreover, the amplitude of this S wave is proportional to the acceleration of the oscillating electron. The intensity of the secondary emission I, however, is proportional to the square of its amplitude. Thus, we arrive at the formula $I \sim \omega^4$.

But if the atmosphere is polluted by emissions, for example, from coal-fired power plants or an active volcano, light is scattered more intensively even during the daytime, and at midday we can see a yellow sky and a red sun.

If our atmosphere had a different chemical composition, the color of the sky and shades of sunsets on the Earth would be different. On Mars, for example, as we already know, the atmosphere is "weak;" its pressure on the planet's surface is about 200 times less than the pressure of the Earth's atmosphere and is made up of almost 95% carbon dioxide. Owing to this, the color of the sky on Mars during the daytime is usually that of butterscotch candy. The reason that the color of the sky on Mars is this way is primarily due to the dustiness of its atmosphere; specifically, the high concentration of black iron oxide in Martian dust. However, when the Sun rises and sets on Mars, the color of the sky changes from pink to light blue around the Sun (Fig. 4.6).

At night, when the sky above us is in the Earth's shadow and the Sun's rays do not penetrate it, radiation dispersion does not occur, which is why the sky appears black to us. Incidentally, this black sky is exactly what astronauts see when they are in orbit because its altitude is higher than that of the Earth's atmosphere. Astronauts are the only ones who can simultaneously see the black sky and the bright white Sun (Fig. 4.7) that to us on the Earth seems so strange and unimaginable because we always see the starry sky from the bottom of the atmosphere.

If the sky is cloudy, then only a small amount of direct sunlight reaches the Earth The light that we see hits our eyes after it has been refracted numerous times through droplets in the air. Each of these many droplets reflects light

Fig. 4.6 A sunset on Mars

Fig. 4.7 A look at the Sun from the Earth's orbit

in different ways and in different directions. In addition, since the size of a droplet is greater than the wavelength of light, different waves from different spectral regions scatter almost the same way. The combination of so much reflected light is what gives us the color white that we see in the sky. When it is overcast and there is dense cloud cover, clouds are high in the sky, which means that a significant amount of sunlight is absorbed and we see a gray sky. Storm clouds are extremely thick and can weaken the intensity of solar radiation more than 1000 times.

4.3   Mirages

The human eye is an optical instrument that captures an image of the outside world on the retina. When interpreting the information that we have received, we usually think a ray of light only travels in a straight line. In actuality, light moves in a straight line only in a homogeneous medium, but the atmosphere is definitely not this type of medium.

> The layers of the atmosphere are made up of different temperatures and densities. Additionally, clouds of water vapor and ice crystals constantly form and change position, which sometimes creates stunning optical displays that distort the images of real objects or make them look as if they exist when in fact they do not. Such phenomena are called *mirages*.

Everyone has most definitely seen a mirage without realizing it. In the hot summer it looks like puddles have suddenly formed on the asphalt in front of a car when in fact nothing is there. If that isn't a mirage, then what is? Who has not seen the Sun set and thought it seemed flattened out both on the top and on the bottom (Fig. 4.8)? But it is most definitely not flattened out; rather, it just seems that way to us.

Fig. 4.8 A sunset at sea

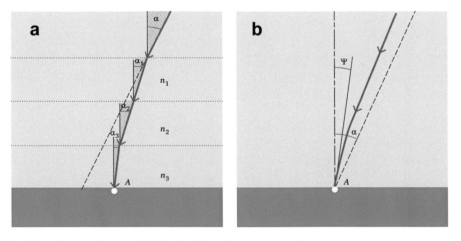

Fig. 4.9 Shape of a ray of light in the planet's atmosphere

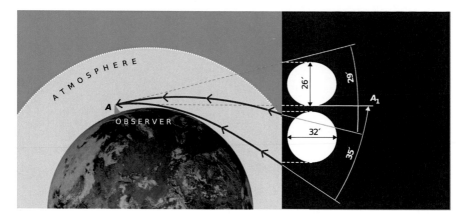

Fig. 4.10   Distortion of the Sun's shape above the horizon

Why the Sun Isn't Round We know that the Earth's atmosphere is similar to a puff pastry pie and air density decreases with increased altitude. The denser the air, the slower the speed of propagation of electromagnetic radiation in the air. If c is the speed of light in a vacuum, and v is the speed of light in a field, then the refractive index of the medium is $n = \frac{c}{v}$. Let's imagine that the atmosphere is made up of several layers with different values of the refractive index. Then a ray of light in the atmosphere will resemble a broken line (Fig. 4.9a).

For all intents and purposes, let's assume that the refractive index increases when moving from the high layers of the atmosphere to the lower ones, i.e.,

$$n_1 > n_2 > n_3. \tag{4.2}$$

By applying Snell's law to our scenario, we can find the ratio between the angles of incidence of beam radiation at the boundaries between atmospheric layers:

$$\frac{\sin \alpha}{\sin \alpha_1} = n_1; \tag{4.3}$$

$$\frac{\sin \alpha_1}{\sin \alpha_2} = \frac{n_2}{n_1}; \tag{4.4}$$

$$\frac{\sin \alpha_2}{\sin \alpha_3} = \frac{n_3}{n_2}. \tag{4.5}$$

Thus, if a ray of light falls on the Earth at the angle α, then because of refraction it will be visible to an observer at the angle ψ. The angle $\Omega = \alpha - \psi$ is called the *angle of refraction*. The boundaries between the layers in the atmosphere are obviously vague, and the refractive index does not change abruptly but gradually. This means that rays of light travel to the atmosphere on a smooth flowing line and not an angled one (Fig. 4.9b). As is evident in Fig. 4.10, the upper limb of the Sun's disk rises up less from refraction than the lower limb and this difference is about 6'. This explains why the Sun is flattened out in a vertical direction.

Fig. 4.11 A visible deformation of the Sun's disk at sunset

The reason that mirages occur is because light rays curve when they pass through the atmosphere. This is called *refraction*. Refraction is due to atmospheric inhomogeneity based on altitude. A ray of light traveling at sunset from the lower limb of the Sun's disk to an onlooker's eye passes through a denser atmospheric layer and bends farther than a ray that travels from the upper limb of the Sun's disk. Therefore, it seems to someone looking at the Sun that its lower limb is slightly raised up as compared to its actual direction.

At noon an observer on the Earth sees the Sun as a disk with a 32′ angular dimension. At sunset this situation is different. When the Sun's lower limb appears to touch the horizon, the size of the horizontal angular dimension remains the same—32′, but the size of the vertical angular dimension decreases to 26′ (Fig. 4.11). When there are more complex atmospheric inhomogeneities, rays of light may bend even more and one can see the Sun split into two flat spots.

4.4 Rainbows

A rainbow is a very beautiful atmospheric phenomenon that everyone loves to see. Rainbows are found in just about every culture's fairy tales, proverbs and superstitious beliefs. For example, there is an amusing legend that buried treasure can be found at the beginning of a rainbow. If you believe this, then there must be two buried treasures and they are buried symmetrically from the point of the observer.

If you try to get close to a rainbow when you see one in order to more accurately figure out where the treasure you are looking for is buried, you will find that its edge moves with you and it will keep on "shooting out" from a new spot.

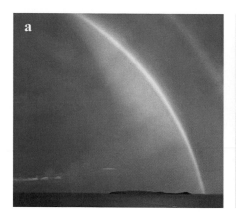

Fig. 4.12 A rainbow: **a** a double rainbow; **b** a rainbow over Victoria Falls (Africa)

But wherever and whenever you see a rainbow, the colors in it will always be in the same order, which is based on the famous mnemonic device "Roy G. Biv" (the sequence of colors in a rainbow from the outer edge is red, orange, yellow, green, blue, indigo and violet).

What is a rainbow (Fig. 4.12)?

> A rainbow is an optical phenomenon. It occurs when a bright light source (such as the Sun or the Moon) illuminates a multitude of drops hanging in the air after rain or fog, for example.

We see a rainbow as an arc of seven colors of the spectrum (actually the colors of a rainbow slowly transition from one shade to another as they move through in-between shades).

The physical reason that rainbows occur is because of drops of water in the atmosphere that reflect and refract sunlight. Now it makes sense why we most often see rainbows after it has rained. Drops deflect light of different colors, i.e., of different wavelengths, at different angles. The refractive index for red, which has a longer wavelength, is less than for violet, which has a short wavelength. For this reason, red light is deflected the least (by 137° 30′), while violet light is deflected the most (by 139° 20′). What we see as a result of this is called a *dispersion of light*, which is the decomposition of sunlight as it passes through a prism (Fig. 4.13).

What conclusions can we make after having closely observed so many rainbows? If an observer is standing with their back to the light source, i.e., the Sun, they see a rainbow as a multi-colored stream of light emanating from

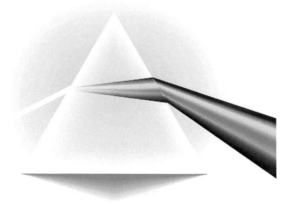

Fig. 4.13 The dispersion of light through a glass prism

a distance along concentric arcs (Fig. 4.14). If you mentally close a rainbow, you have a circle. Its center lies on a straight line that passes through the observer and the Sun, which is always behind the observer; however, it is impossible to see the Sun and a rainbow at the same time. The angular radius of a circle's periphery is 42°. The lower the Sun is above the horizon, the closer the arc of the rainbow is to half of the circle, while the height of the top of its arc, on the other hand, is closer to 42° above the Earth's surface. The most that we can see of a rainbow on the Earth is a semicircle, which becomes visible when the Sun meets the horizon. But if you climb a mountain or fly in a plane, you can see the full circle of a rainbow; in other words, the higher the point where you are, the fuller the rainbow. A rainbow cannot be seen if the Sun is higher than 42° above the horizon because then a rainbow's circumference is below ground level.

In sunny weather, you can always make your own little artificial rainbow. In order to do this, you need to be in a direction opposite the Sun and spray water so that many droplets form (you can do this however possible, for example, with a garden sprinkler). After a short time, while the drops are hanging in the air, you can see a little artificial rainbow.

The features of a specific rainbow—the brightness of its colors, the width of its bands, how long it is visible—depend on the number of drops per unit volume of air and their size. If the drops that make up a rainbow were in the atmosphere indefinitely, the maximum amount of time that it would remain

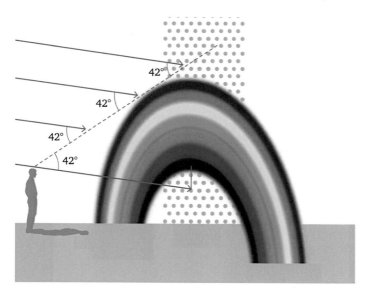

Fig. 4.14 The conditions that must be met in order to see a rainbow

visible would be 2 h and 48 min. This is exactly the amount of time it takes for the Sun to pass through the sky on a path that corresponds to the length of a rainbow's arc with an angular dimension of 42°. But drops in the air do not "live" long: either they evaporate or they begin to move downward once they have joined together and increased in mass. When the number of drops decreases, a rainbow disappears.

Usually when we see a rainbow, red is on the outer edge of the arc and violet is on the inner edge. This type of rainbow is called a *primary rainbow*. As we will see, when a primary rainbow is forming, light undergoes a total internal reflection in a drop of water. Occasionally, a faint secondary rainbow appears simultaneously with the primary one (Fig. 4.15). A secondary rainbow is formed by rays of light that are reflected in drops twice. In this rainbow, colors are in the reverse order: violet is on the outer edge of the arc and red is on the inner edge. The angular radius of a secondary rainbow is 53°. The color of the sky between the two rainbows seems to be darker. In theory, a third rainbow could also appear where light undergoes three reflections inside of drops, but in a natural environment this happens extremely seldom.

Now let's look at the scientific process behind the formation of a rainbow and try to understand all of the aforementioned nuances of this remarkable optical phenomenon. In Fig. 4.16, point O denotes the position of an observer, point O_1 is the point opposite of the Sun's direction and is located

Fig. 4.15 A rainbow in the Rocky Mountains (Banff National Park, Canada)

below the horizon, OCD is the Earth's surface plane, angle $\widehat{AOO_1} = \psi$ is the solar altitude angle above the horizon and segment CD is the skyline.

The observer is at the vertex of a cone with an axis that joins the observer and the Sun, while the rainbow is part of the circumference that makes up the base of the cone, which is located above the horizon. The observer sees only that part of the rainbow that is above the horizon (CBD arc). The observer can see the top of the rainbow at angle $\Phi = \widehat{AOB}$ and the base is visible at

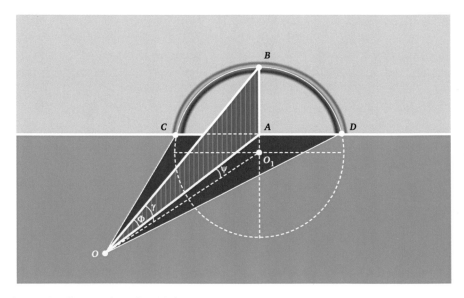

Fig. 4.16 Geometrics of a rainbow

angle $\alpha = \widehat{AOD}$. The angle of altitude of the rainbow $\gamma = \psi + \Phi$ depends on where the observer is located in relation to the Sun, and the size of the rainbow is determined by the height of the Sun above the horizon (angle ψ can be easily found by measuring the height of the observer and the length of the shadow they cast). When the observer moves, the cone will also move. It is precisely for this reason that it is impossible to reach a rainbow.

Rainbow Observation Parameters We will consider the reflection and refraction of a ray of light that passes through a drop of water.

To do this, we need to bring in one additional value—the impact parameter ξ:

$$\xi = \frac{\rho}{R} \tag{4.6}$$

where R is the radius of the drop and ρ is the distance from the axis of the ray to a straight line parallel to it that passes through the center of the drop (Fig. 4.17).

Let's assume that a parallel cone of light rays falls on the drop.

Because of spherical symmetry, all of the light rays that fall on the drop with the same value ξ will move inside of it along symmetrical paths and exit it at the same angle to go toward their original direction. The trajectory of each ray lies on a plane that contains the light rays' initial path and a straight

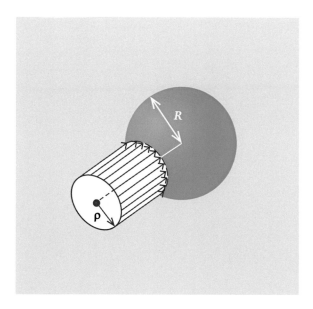

Fig. 4.17 The impact parameter

line parallel to it drawn through the center of the drop. The movement of a light ray with the impact parameter ξ on this plane is shown in Fig. 4.18. It is important to note that we are not omitting from consideration those light rays that are reflected at points A and C and refracted at point B.

If α is the solar incidence angle on the surface of a drop, then $\sin \alpha = \xi$ (see Fig. 4.18). We have: $\widehat{OAB} = \widehat{ABO}$ and $\widehat{OBC} = \widehat{BCO}$. At the same time, $\widehat{ABO} = \widehat{OBC}$. These four angles, which are equal to each other, we denote as β. The line OO' is the axis of symmetry for the trajectory of the light ray. Therefore, at points A and C the refracted ray turns through an angle of $\alpha - \beta$, while at point C it turns through an angle of $\pi - 2\beta$. The light ray that is exiting the droplet in the direction CC_1 will be turned with regard to the original direction through an of angle $2(\alpha - \beta) + (\pi - 2\beta) = \pi + 2\alpha - 4\beta$. But this angle is equal to $\pi - \varphi$. Thus,

$$\varphi = 4\beta - 2\alpha. \tag{4.7}$$

We will find the connection between the angle φ and the impact parameter of the ray. At point A, we have $\frac{\sin \alpha}{\sin \beta} = n$ (n is the refractive index of water). We get:

$$\frac{\sin \alpha}{n} = \sin \frac{\varphi + 2\alpha}{4}; \tag{4.8}$$

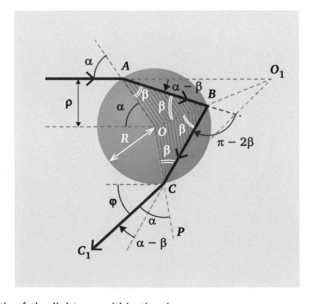

Fig. 4.18 Path of the light ray within the drop

$$\frac{\varphi + 2\alpha}{4} = \arcsin\frac{\sin\alpha}{n}. \tag{4.9}$$

Considering that $\sin\alpha = \xi$, we can write down:

$$\varphi = 4\arcsin\frac{\xi}{n} - 2\arcsin\xi. \tag{4.10}$$

The dependency graph of φ and ξ is shown in Fig. 4.19. As the impact parameter grows, the angle φ increases up to a certain point and then begins to decrease as it reverts to 0.

We will consider the critical implications of this formula. First, we will find the value of the impact parameter at which a ray of light exits a droplet in the direction that is opposite to its initial direction. This means that $\gamma = 0$. We have

$$2\arcsin\frac{\xi}{n} = \arcsin\xi. \tag{4.11}$$

Or

$$\xi = \sin\left(2\arcsin\frac{\xi}{n}\right). \tag{4.12}$$

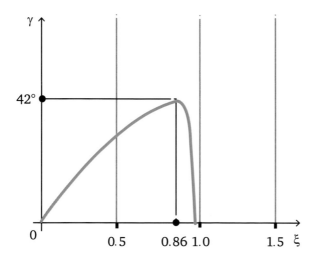

Fig. 4.19 Dependence of the refraction angle on the impact parameter

Using the trigonometric formula $\sin 2x = 2 \sin x \times \cos x = 2 \sin x \times \sqrt{1 - (\sin x)^2}$, we get:

$$\xi = \frac{n}{2}\sqrt{4 - n^2}. \tag{4.13}$$

If $n = \frac{4}{3}$ (this is a good estimate for the visible spectrum), then $\xi = 0.996$.

Now we will determine the values of ξ at which the angle between falling and emergent light rays has the highest possible value. In order to do this, we find the extreme point of our coversine. Its differential coefficient is

$$\frac{d\gamma}{d\xi} = \frac{4}{\sqrt{n^2 - \xi^2}} - \frac{2}{\sqrt{1 - \xi^2}}. \tag{4.14}$$

When it is equal to zero, we get the equation:

$$\sqrt{n^2 - \xi^2} = 2\sqrt{1 - \xi^2}. \tag{4.15}$$

Its solution gives us

$$\xi_0 = \sqrt{\frac{4 - n^2}{3}} = 0.861. \tag{4.16}$$

If we substitute this value and $n = \frac{4}{3}$ into the formula for angle γ, we find that the maximum possible value of the angle is $\gamma_{max} = 42° \, 02'$.

In summary, light rays having all possible values of ξ from zero to one fall on the drop to be reflected and refracted at different angles of γ. An observer will see the brightest rays that have a minimum divergence, that is, those that fall in the area of the curve peak as in Fig. 4.19 (i.e., $\gamma \approx \gamma_{max}$).

Now we will explain why a rainbow is made up of a multicolored arc. For simplicity's sake, we will consider a ray of light that consists of a mixture of two wavelengths—red and violet rays. The radiation they emit has different refractive indexes: $n_{red} = 1.331$ and $n_{violet} = 1.344$. According to our formulas, we find $\xi_{red} = 0.862$, $\gamma_{red} = 42° \, 22'$ and $\xi_{violet} = 0.855$, $\gamma_{violet} = 40° \, 36'$. The path of the red and violet rays is shown in Fig. 4.20. It is clear that when someone is looking at a rainbow, the red arc will be visible at an angle of $42° \, 22'$ and the violet arc will be seen at an angle of $40° \, 36'$. This explains why rainbows are multicolored and their outer edge is red and their inner edge is violet. The same type of calculation applies to the in-between colors of the rainbow.

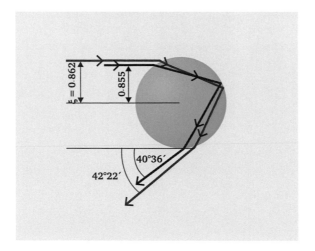

Fig. 4.20 Path of the red and violet rays within the drop

4.5 Wind

What makes the wind blow? It is definitely not because trees sway and move the air! The answer is simple: the wind blows due to the movement of air masses from high to low pressure areas. In a room where there is a heater, for example, the air above it warms up, rises to the ceiling, cools down and then sinks along the opposite wall. There is continuous air circulation. This theory is directly applicable to the movement of the Earth's atmospheric air masses. In this case, the "heater" is the equator region where the Sun's rays fall almost vertically and warm up the air, which rises up to a height of 15–17 km (9.32–10.56 mi). The direction of air currents by the Equator is parallel to the Earth's surface: north of the Equator air begins to move to the north, while south of it air moves to the south. At the same time, air masses cool down, as a result of which they descend at the middle latitudes. Thereafter, air moves back to the Equator (Fig. 4.21).

The main reason that there is wind on the Earth is because of equatorial convection. The average wind speed near the Earth's surface is about 10 m (32.08 ft)/s. Wind speed depends on how high the wind is above the surface of the Earth: the higher it is, the greater its speed. However, the direction of true wind differs from that shown in Fig. 4.21. In both the Northern and Southern Hemispheres, surface winds do not blow in the direction of the Equator but rather change direction and become easterly; in other words, they blow from east to west. Conversely, upper-level winds in the upper troposphere become westerly. The reason that wind changes direction is the rotation of the Earth and the Coriolis force that is related to it.

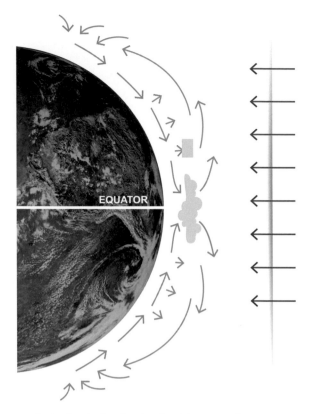

Fig. 4.21 Scheme of air circulation in the Earth's atmosphere

The Average Speed of Wind on the Earth We will estimate the average wind speed near the Earth's surface. We already know that the energy from the solar radiation that reaches the Earth's surface is about 10^{17} W. In actuality, only about 1% of this energy becomes mechanical wind energy; the other 99% is transferred to infrared radiation for the Earth. Thus, the mechanical power that comes from wind is approximately 10^{15} W. The kinetic energy from it is $\frac{m_a v^2}{2}$ where v is average wind speed and m_a is atmospheric mass. The kinetic energy from wind is equal to power multiplied by time τ, which is the characteristic time during which an air mass goes around the Earth. We get: $v \approx 10\,\text{m}\,(32.80\,\text{ft})/\text{s}$, $\tau \approx 1$ week.
. The first assessment provides the approximate wind speed near the Earth's surface. When moving away from it (i.e., when ascending), wind strength significantly increases. The reason for this is because of air mass balance, which is based on the constancy of flow of matter near the surface and at an elevation: $(v\rho)_{\text{surf}} = (v\rho)_{\text{elev}}$. Since the atmospheric density ρ is much lower at elevations of 15–17 km (9.32–10.56 mi) than near the Earth's surface, wind speed is correspondingly much greater—up to 100 m (328 ft)/s and more. The speed of movement of air masses that travel from the Equator to the north and south in the upper troposphere is about 200 m (656.17 ft)/s!

The Coriolis force is close to zero near the Equator, and air currents retain their original direction of motion. Further away from this area, a deflection of air currents begins. A 90° rotation occurs at a certain latitude; in both hemispheres, the flow of upper-level winds shifts from west to east, while the flow of low-level winds changes from east to west. This latitude is determined by the equation

$$\operatorname{tg} \varphi = \frac{v}{\omega_E R_E}, \varphi \approx 30°. \tag{4.17}$$

Consequently, in the Northern and Southern Hemispheres two powerful high-altitude jet streams—the northern jet stream and the southern jet stream—flow from west to east (Fig. 4.22). The average height of these regions of air (also known as polar vortices) is 9–10 km (5.59–6.21 mi), and their wind speed is 30–35 km (18.64–21.75 mi)/h. It takes 8–10 days for this air to go around the Earth. Travelers who try to complete a round-the-world hot air balloon ride—most often in the Southern Hemisphere—find these bands of strong wind much to their advantage. Taking a similar trip in the Northern Hemisphere is difficult because of the Himalayan Mountains.

When circumpolar currents pass over land, they flow closer to the Equator, but they change position over oceans and move to higher latitudes because

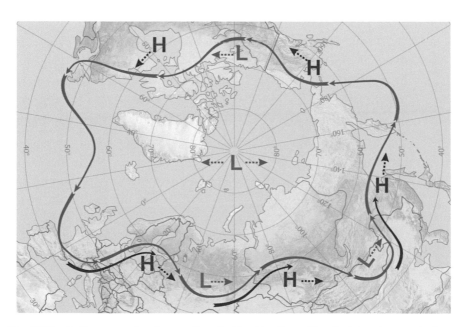

Fig. 4.22 A diagram of polar vortices in the Northern Hemisphere. Letters indicate areas of low (L) and high (H) pressure

convection over oceans is more intensive. First, this is due to the fact that more solar energy is absorbed over large bodies of water than over land since the albedo of the ocean surface is lower than the albedo of land. In addition, humidity is higher over the oceans. For this reason, convective lifting occurs at a higher altitude over the oceans than over land, and, consequently, as air masses move away from the Equator, they travel at a higher speed. The location of polar vortices often changes and they randomly bend, which is one of the main reasons why the weather changes.

How Air Pressure Changes with Altitude We will first obtain an equation for the change in atmospheric pressure with altitude. We know the equation for the ideal gas law very well. If we express the volume of gas in terms of its density ρ, we get:

$$p = \frac{N_A k_B T}{\mu}\rho. \tag{4.18}$$

How does pressure depend on altitude? If we were working with a liquid, we would know the answer: $p = \rho g z$. But in a liquid, density is constant; however, in the atmosphere, as we already know, density changes with altitude. Therefore, we can write a similar formula not for p and z, but only for their infinitesimally small changes dp and dz:

$$dp = -\rho g dz. \tag{4.19}$$

The minus sign indicates that atmospheric pressure decreases with altitude. By combining these two formulas, we obtain an equation for the ratio of the infinitesimal changes in pressure and altitude $\frac{dp}{dz}$:

$$\frac{1}{p}\frac{dp}{dz} = -\frac{\mu g}{N_A k_B T}. \tag{4.20}$$

When convective lifting occurs, air masses expand as they move into rarefied atmospheric layers. Because these masses expand in low pressure areas, air cools down. When the temperature drops below the condensation point (i.e., the dew point), clouds form. For this reason, the sky near the Equator is almost always covered with dense clouds.

The air in the tropics rises to 17 km (10.56 mi), although its temperature drops to − 75 °C (− 103 °F) and the humidity level falls to almost zero because water vapor remains in the clouds.

Near the Tropic of Cancer and the Tropic of Capricorn, polar vortices cause air to descend to the Earth's surface. Because of solar energy, air masses close to the Equator accumulate a large amount of internal energy, which is used to lift air up to high altitudes. When the amount of this energy decreases in

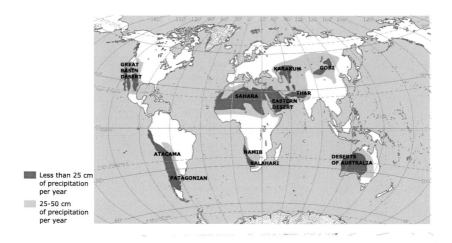

Fig. 4.23   Location of the largest deserts on the world map

the tropics between the latitudes of 25 and 30°, it is released back and heats up the air. Hence, this air becomes very warm (up to 30 °C [86 °F]) at the Earth's surface and—unlike at the Equator—dry. For this reason, the largest deserts are located in these very latitudes (Fig. 4.23).

In certain areas in the ocean, one finds calm seas due to the fact that the air there blows at a low horizontal speed. These areas are found under polar vortices where hot air descends. In ancient times, these windless areas (between 25° north and 30° south of the Equator) came to be known by sailors as the *horse latitudes*.

> In the days that sailing ships transported horses from Europe to America, crews were unable to sail when there was a lack of wind and, thus, they would run out of drinking water. When this happened, they were forced to throw their horses overboard in order to converse water. This is why the subtropical latitudes are referred to as the *horse latitudes*.

The main difference between the air in the horse latitudes and by the Equator is low humidity. Thus, equatorial air masses saturated with water vapor have a lower molar mass than the air masses in the horse latitudes because $\mu_{H_2O} < \mu_{air}$.

Since the air at the Equator is lighter, when the pressure there rises, it falls more slowly than in the horse latitudes where the air is dry. Hence, the coldest air in the troposphere is over the Equator where it drops down to − 75 °C (−103 °F), while the air over the horse latitudes is 10–15 °C (50–59 °F) warmer. This is the way that the system of high- and low-pressure zones in the atmosphere acts. The former is located on both sides of the Equator at latitudes of approximately 35° and at the poles at latitudes above 65°, while the latter is found along the Equator and in subpolar latitudes (Fig. 4.24).

The lack of homogeneity in atmospheric pressure results in steady air currents, with upper- and lower-level winds moving in opposite directions. The Coriolis force acts on the initial direction of these currents. It deflects air currents in a clockwise direction in the Northern Hemisphere and in a counter-clockwise direction in the Southern Hemisphere. The resulting northeasterly winds of the Northern Hemisphere and southeasterly winds of the Southern Hemisphere are called *trade winds*. When trade winds are close to the Equator, they blow practically parallel to it. This type of tropical atmospheric cell of circulation is called the *Hadley cell* (Fig. 4.25).

When air currents are forming trade winds, some of the air that is beneath polar vortices flows away from the Equator, as well as toward it, thus creating another atmospheric circulation cell—the *Ferrel cell*. This type of circulation occurs between the latitudes of 30° and 65° and the direction of air movement is opposite to that which occurs in a tropical circulation cell. At these latitudes, the Coriolis force deflects the winds in the opposite direction. Thus,

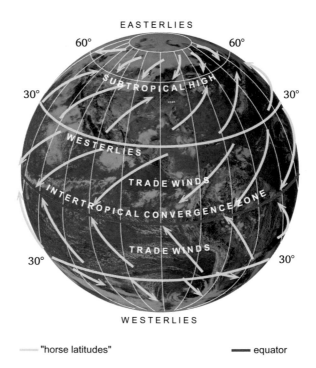

Fig. 4.24 The main atmospheric circulation patterns on the Earth

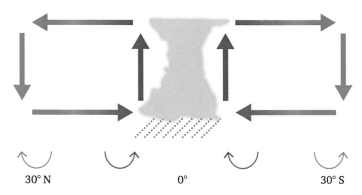

Fig. 4.25 Emergence of Hadley cells

at the middle latitudes, westerly winds prevail. In the Northern Hemisphere, mountain ranges weaken the strength of these winds. In the Southern Hemisphere, the Andes are not tall enough to block them and as a result, these winds pick up speed over the ocean and have the strength of a hurricane. Sailors call this area of strong westerly winds the *Roaring Forties* and took advantage of them when sailing from Europe to Australia (Fig. 4.26).

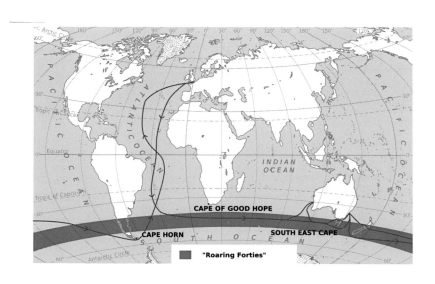

Fig. 4.26 A typical sailing route around the world from Europe to Australia

> The zone located between the area where the trade winds of the Northern and Southern Hemispheres converge is known as the *Intertropical Convergence Zone.*

During the course of a year, this zone can shift from the Equator to a warmer hemisphere that is experiencing summer at that time. Therefore, if in a specific place in winter the primary direction of wind reversal is from west to east, in summer it changes and winds blow from east to west. This is especially evident in the Indian Ocean basin. This type of wind reversal is known as a *tropical monsoon.*

Thus, monsoons are seasonal winds that blow annually in tropical regions for a certain number of months. Monsoon winds bring with them heavy rainfall and blow northeast over vast expanses of the Indian Ocean. In particular, they affect the Indian subcontinent and the Southeast Asian Peninsula, South China, the islands of Indonesia, the latitudes of the southern part of the Indian Ocean to the north of Australia and Madagascar, as well as large parts of Africa. The direction of monsoons' movement is determined by low-pressure areas that form over North America and Africa from May to July and over Australia in December and January because of an increase in air temperature over the tropics in the summer months. The particularly strong development of monsoons in the Indian Ocean is due both to the specific influence exerted by the large continent Eurasia to the north, and to the fact that Africa, which is stretched out onto both the Northern and Southern Hemispheres, is very close to it.

A constant exchange of warm and cold air occurs between the tropical air circulation zone and the mid-latitude zone. As a result, a transfer of warm air from low to high latitudes and of cold air from high to low latitudes takes place. It is precisely this process that helps maintain the Earth's thermal equilibrium.

Closer to the North and South Poles, the direction of air circulation changes again. Air rises where there is low pressure and falls at the poles. Thanks to the Coriolis force, easterly winds prevail in the polar regions, which results in the development of polar cells of atmospheric circulation (Fig. 4.27).

Figure 4.28 provides an overall picture of what atmospheric winds on the Earth look like.

In addition to global winds, there are also local winds that are specific solely to the geographical areas in which they are located and are caused by unique geographical and climatic features. There are different types of natural elements and forces responsible for making these winds blow.

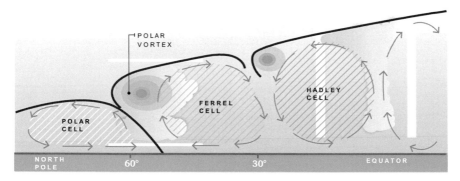

Fig. 4.27 Components of atmospheric circulation in the Northern Hemisphere

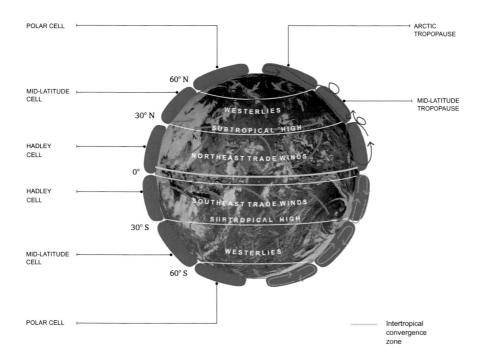

Fig. 4.28 General circulation of the Earth's atmosphere

First, local winds may develop because of air circulation in an area that does not depend on the general air circulation in the atmosphere. These types of local winds may include breezes, i.e., wind that blows on the shores of seas and large lakes. The physical reason that breezes occur is because heat is not equally distributed between the seashore and the water in the day and at night. Hence, during the day it blows from the cool water onto the land, which has heated up, while at night, on the other hand, it blows from the

Fig. 4.29 Formation of a breeze: **a** day sea breeze; **b** night coastal breeze

surface of the cool land onto the sea, which has retained the day's warmth. The formation of a breeze is shown in Fig. 4.29. One can see that in the day sea breezes blow from the water toward a landmass (Fig. 4.29a), while at night land breezes blow away from it and toward the water (Fig. 4.29b).

Second, local winds are caused by surface topography. When we refer to local winds, we mean mountain-valley winds that blow daily and therefore seem just like a breeze. During the day, valley winds transfer air from the lower areas in valleys upward and into mountains. However, at night this process is reversed and mountain winds descend down slopes and valleys. An example of a type of mountain-valley wind is foehn wind. This is a warm wind that blows down mountain slopes into valleys when an air current crosses over a chain of mountains. The downward movement of foehn wind leads to an increase in air temperature. This occurs because mountains have an effect on atmospheric circulation (Fig. 4.30). An air current produces the first oscillation (A), which happens again after it passes over a mountain (B). As a result of this, distinctive lenticular clouds form at a mountain's highest points (Fig. 4.31). These clouds are unique in that they do not move and are practically "attached" to mountain tops no matter how strong the wind is.

Topography also affects the formation of a type of wind called *bora*. It is a northerly wind, which, while crossing over mountains, moves downward and intensifies, sometimes quite dramatically. Bora is found, for example in Gelendzhik Bay and Novorossiysk Bay in Russia. It also exists on Lake Baikal where it is known as *Sarma*. It is often the case that the characteristics unique to local winds are caused by air passing over a dry surface that has been warmed up, for example, a desert, or, conversely, over a rapidly evaporating water surface such as a large lake.

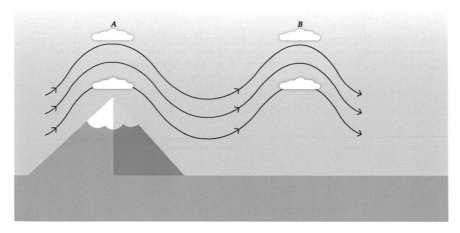

Fig. 4.30 Development of foehn wind and lenticular clouds

Fig. 4.31 Lenticular clouds over mountains near Queenstown (New Zealand)

Third and lastly, local winds develop due to the flow of general air circulation, which is specific in nature due to unique aspects of different areas. For example, in the Mediterranean Sea sirocco winds cause dusty dry conditions, while in some parts of Europe they bring heat and humidity. Other localized winds that have specific names are well known in various places on the Earth.

It often happens that areas of reduced air pressure appear over certain places on the Earth's surface. Because of a difference in pressure, air currents (i.e., winds) from high-pressure areas rush into those areas where there is low pressure. The result is the formation of a cyclone, in the center of which air rises up, then cools down and clouds form (Fig. 4.32a). The Coriolis

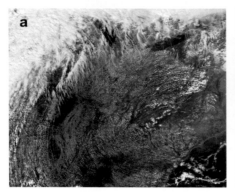 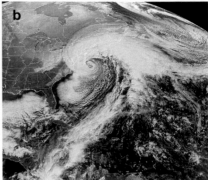

Fig. 4.32 Pictures taken from a spaceship: **a** cyclone; **b** anticyclone

force causes surface winds in the clouds' path to swirl. The air pressure in the upper part of the cyclone, however, is higher than the average pressure at that altitude. For this reason, the direction of air movement there comes from the cyclone's center. An anticyclone, on the other hand, forms when there is an area of high pressure close to the Earth's surface (Fig. 4.32b). In this case, dry (and usually cold) air descends from the troposphere and swirls under the Coriolis force. The pressure in that area of the troposphere decreases in relation to the average amount of pressure for the given altitude and air then rushes in from adjacent areas of the atmosphere. When there is an anticyclone, clouds do not form and the sky is clear.

Cyclones and anticyclones "live" in the atmosphere for about one week and their dimensions range from 500 to 3000 km (310.69–1864.11 mi).

Therefore, we know that wind is a very complex concept affected by a variety of factors ranging from global processes that warm up the Earth's surface at the Equator to local ones associated with the fact that, for example, a local pond does not have a chance to cool down overnight. The fact that wind is so complex is wonderful because if it were completely predictable, it would be of little interest to anyone.

Further Reading

1 Byalko, A.V.: Our Planet the Earth. MIR Publisher, Moscow (1983)
2 Kaimal, J.C., Finnigan, J.J.: Atmospheric Boundary Layer Flows: Their Structure and Measurement. Oxford University Press, Oxford (1998)
3 Lewin, W., Goldstein, W.: For the Love of Physics. Taxmann Publications (2012)
4 Petty, G.W.: A First Course in Atmospheric Radiation, 2nd edn. Sundog Publishing (2006)

5 Ralph, F.M., Dettinger, M.D., Rutz, J.J., Waliser, D.E.: Atmospheric Rivers. Springer, Berlin (2020)
6 Varlamov, A.A., Aslamazov, L.G.: The Wonders of Physics, 4th edn. World Scientific, Singapore (2019)
7 Wallace, J.M., Hobbs P.V.: Atmospheric Science: An Introductory Survey, 2nd edn. Academic Press, Cambridge (2006)

5

The Ocean

Abstract In this chapter, we discuss the physical phenomena related to the existence of the Ocean on the Earth's surface. The chemical composition and mass of water in oceans and seas on the Earth are considered. Then we study the main features of ocean currents. Additionally, we review the physics of oceanic wave formation and the various types of waves in the Ocean. Such phenomena as foams on sea waves, sea swells, tsunamis, moonglades, and the color of the sea depending on weather conditions are discussed.

If we look at the Earth from outer space, we cannot help but ask the question: Why is our planet called "Earth" and not "Water" or "Ocean"? After all, the World Ocean makes up more than 70% of the area of our planet! Moreover, there are seas, rivers and glaciers on the continents that take up additional expanses of land. But if we compare areas by mass, the mass of water ($M_{\mathrm{M.oc}} = 1.4 \times 10^{21}$ kg) is only 1/4400 part of the Earth's mass. For this reason, naming our planet "Earth" is well justified. If the surface of the Earth were a regular sphere, the ocean would have the same depth at any point and be equal to:

$$\frac{M_{\mathrm{M.oc}}}{4\pi\,\mathrm{R}_{\mathrm{E}}^2 \rho_{\mathrm{w}}} = 2.75\,\mathrm{km}\ (1.71\,\mathrm{mi}) \tag{5.1}$$

© The Author(s), under exclusive license to Springer Nature Switzerland AG 2023
D. Livanov, *The Physics of Planet Earth and Its Natural Wonders*,
https://doi.org/10.1007/978-3-031-33426-9_5

Table 5.1 Data for the constituent parts of the World Oceans

Ocean	Area, $\times 10^6$ km^2	Average depth, m (ft)	Maximum depth, m (ft)
Pacific Ocean	180	4028 (13,215)	11,971 (39,274)
Atlantic Ocean	107	3332 (10,931)	9219 (30,246)
Indian Ocean	74	3897 (12,785)	7455 (24,458)
Arctic Ocean	15	1225 (4019)	5625 (18,454)
Southern Ocean	20	3503 (11,492)	7235 (23,736)

Compared to the radius of the Earth (6400 km [3976 mi]), that is a very thin layer and akin to simply moistening a stone with water.

Today geographers recognize five oceans: the Pacific, Atlantic, Indian, Arctic and Southern. In 2000, the International Hydrographic Organization "removed" the southern sections of the Atlantic, Indian and Pacific Oceans to delimit the Southern Ocean. The reason this was done is because in that particular area of the Southern Ocean the circumpolar current, which sets its own rules for the weather patterns that develop, has caused a unique water flow pattern to appear. Therefore, the Southern Ocean should most definitely not be disregarded, but perhaps should be regarded even higher than the other oceans because they, unlike the Southern Ocean, have very vague and artificially determined borders between them.

Baseline data for the constituent parts of the World Oceans are shown in Table 5.1.

5.1 Special Qualities of Sea and Ocean Water

Any one of us can immediately answer the question: What makes seawater different from lake and river water? Seawater is salty and denser than fresh water because of the chemical compounds that dissolve in it. If we examine this fact according to the rules of chemistry, sea and ocean water is a weak solution that contains approximately 4% salt. Seawater has all of the characteristics of weak solutions; the most important one for us is that it has a low freezing point in relation to pure water.

Oceans have been on the Earth slightly less than it has existed, and during that time almost all of the soluble chemical compounds—mainly salt—that were found on the Earth have dissolved in oceans. We know that salts in water dissociate into ions. The main positively-charged ions are: Na^{+1} (30.6%), Mg^{+2} (3.7%), Ca^{+2} (1.2%), K^{+1} (1.1%), and the main negatively-charged ions are: Cl^{-1} (55.0%), SO_4^{-2} (7.7%), HCO_3^{-1} (0.4%) and Br^{-1} (0.2%).

There are 35.2 g (0.078 lbs.) of salt in 1 kg (2.20 lbs.) of seawater. This measurement of value is called *salinity* and the amount of it in seawater can vary from place to place. The main reason for this variation is that water actively evaporates, which therefore increases salinity. Heavy rain and the confluence of rivers, however, lower the salt content in water. Thus, the salinity of ocean water varies depending on geographic latitude. As is evident in the graph (Fig. 5.1), the highest level of salinity is found in the northern and southern latitudes between 25 and 30°.

This is due to the fact that in these latitudes, water evaporates quickly and precipitation is low. Salinity decreases when moving towards higher latitudes and the Equator.

In addition, gases that have dissolved in the Earth's atmosphere are found near the surface of the ocean. Carbon dioxide is the gas that exists in greatest abundance there and it is 5×10^{-4} of the mass of water. There is slightly less nitrogen in the ocean—10^{-5} of the mass of water—and even less oxygen—only 8×10^{-6} of the mass of water. The amount of CO_2 in the ocean is about 60 times higher than the amount of carbon dioxide in the atmosphere, but 130 times less oxygen is dissolved in the ocean than is found in the atmosphere.

Besides salt and gas, organic matter, biogenic substances (i.e., the so-called waste products of living organisms and remains of dead organisms) and microelements also make up the composition of the ocean.

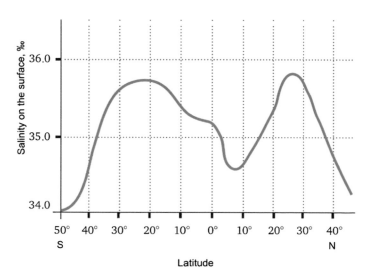

Fig. 5.1 Salinity of the World Ocean depending on latitude

Fig. 5.2 Production of sea salt

For example, in 1^3 km of seawater, there is 2500 T (2755 sh. tn.) of silver and even 50 kg (110 lbs.) of gold!

Sea water is not suitable for drinking or watering plants. For many centuries obtaining clean drinking water was the biggest problem that sailors faced because technology to desalinate water had not yet been developed. This is no longer a problem now and entire countries, for example, ones on the Arabian Peninsula, obtain fresh water thanks to desalination. Sea salt is added to food. Unlike rock salt that is mined from shafts, sea salt is produced through evaporation at special facilities located by natural bays or lakes filled with seawater (Fig. 5.2). Exposure to sun and wind makes the water in these reservoirs evaporate and the salt that remains (after other chemical compounds have been separated and removed) is collected and packaged.

Thus, seawater is a mixture of mineral and organic substances in different forms and states. It is important that the ocean is in a state of chemical equilibrium with the Earth's atmosphere and crust.

5.2   Ocean Currents

We are used to the fact that there is a current in the river, but when we look at the sea, it does not seem at all as though there is a current there as well. In actuality, there are currents in the sea and very large ones too in fact! People who spend a significant amount of time on boats such as sailors, fishermen and yachtsmen know this better than anyone else. The height of currents in the ocean and the sea is one hundred times greater than that of river currents. In addition, the function of these currents is invaluable because thanks to them, heat is transferred from the Equator to higher latitudes. If, for example, a current such as the Gulf Stream disappeared, all of northern Europe would freeze and turn into a glacier. The volume of the Gulf Stream reaches 106 hm^3 (4 billion cubic feet) of water per second, which is about 100 times more than the total runoff from all of the rivers on the planet. The width of the Gulf Stream is 110–120 km (68.35–74.56 mi) and its depth is 700–800 m (2297–2624.67 ft).

However, the Gulf Stream is just one of the currents that flows (Fig. 5.3).

Sea currents are caused by two forces—friction and gravity—but the main source of energy that drives them is wind.

We have already mentioned that the salinity of ocean water can vary greatly from place to place due, on the one hand, to uneven evaporation processes and, on the other, to rainfall and glacier melting. The density of salt water is greater than that of fresh water. If, for example, the air temperature is 15 °C (59 °F), seawater with average salinity has a density of 1.026 g/cm^3. Its density increases, albeit only slightly, with depth and an increase in pressure.

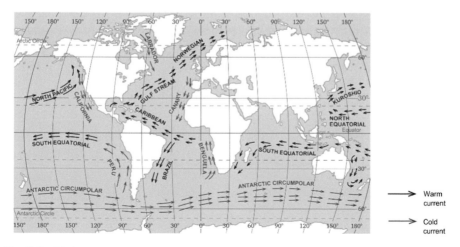

Fig. 5.3 Major ocean currents

If in a certain area of the World Ocean air that is hot and dry due to lack of rainfall causes the surface layer of water to actively evaporate and salinize, then water becomes denser on the surface than it does deep below it. Because of this, a current begins to flow, which is the salty water sinking to the depths of the ocean. This is precisely the way that bodies of water move, for example, in the Mediterranean Sea. A large body of ocean water with an average level of ocean salinity flows into the narrow Strait of Gibraltar. As water moves to the east, due to active evaporation its salinity level increases and reaches 40% off the coast of Turkey. At the same time, water that is salty and heavier sinks to the deep layers of the ocean where a current forms in the bottom layer that flows to the west.

But the main factor that causes ocean currents has to do with the wind, which not only propels sailboats but also affects water. When an area is large, this effect is very significant, so much so that if the wind blows long enough, it begins to drag the top layers of ocean water with it.

Surprisingly, the wind can cause a current that is deep within the ocean's depths to flow in an opposite direction because of the Earth's rotation! The Swedish scientist Vagn Ekman developed the theory of wind-driven currents in the open ocean by taking into account the influence of the Coriolis effect. The strength of this force together with the influence exerted by wind cause deep ocean currents to bend. Let's suppose that we are in the Northern Hemisphere. Then the surface current that the wind caused will deviate to the right of the wind's direction. However, in the Southern Hemisphere the opposite is true—the surface current deviates to the left of its direction. Because of friction, the surface layer of water sets the underlying water layer in motion, which also deviates to the right, and so on. With increased bending and depth, the current gradually becomes weaker. If the wind speed is 10 m (32.81 ft)/s, then the speed of the current on the ocean's surface will be about 0.1 m (0.33 ft)/s, but the usual depth at which the current changes its course and flows in the opposite direction is about 100 m (328 ft). Imagine that at such a relatively shallow depth as 100 m (328 ft) ocean currents reverse their direction!

Wind Speed of an Ocean Current Let's say the wind blows over the ocean at the speed V. The air that is moving because of the wind experiences friction against the water's surface, thus creating tangential stress on it. Tangential stress σ is the friction force divided by area and has the dimensionality $\frac{N}{m^2} = \frac{kg}{m\,s^2}$. The shear stress in water that is caused by wind only depends on wind speed V and air density ρ_{air}. Only the combination of V and ρ_{air}, which has the required dimensionality $\frac{kg}{m\,s^2}$, is $\rho_{air} V^2$. The constant of proportionality was measured experimentally and it turned out to be 2.5×10^{-3}.

The pressure in the surface layer of water causes the liquid mass to move, while the speed of the current v will depend on the depth z. The speed decreases with depth because viscosity impedes the movement of the water layers. One can write:

$$\sigma = -\eta \frac{dv}{dz}, \tag{5.2}$$

where η is viscosity. The minus sign in the previous formula takes into account the fact that the viscous force and the speed are contra-directional.

Let's assume that at a certain depth H, the speed of the current reduces to zero and the speed of the current at the surface itself is v_0. Then according to the linear law, we get that the speed will decrease with increasing depth:

$$v = v_0 \left(1 - \frac{z}{H}\right); \quad v_0 = \frac{\sigma H}{\eta}. \tag{5.3}$$

If we substitute $H = 1\,\text{m}$, $V = 1\,\text{m/s}$, $\rho_{air} = 1.2\,\text{kg/m}^3$ and the viscosity of water $\eta_{water} = 10^{-3}\,\frac{\text{kg}}{\text{m s}}$ into this formula, we get $v_0 = 3\,\text{m}\,(9.84\,\text{ft})/\text{s}$. The reason for this error is that we considered the flow of a liquid as if it were parallel to the plane (i.e., laminar). In fact, this is the case only at very low speeds, but in reality, the layers of the liquid begin to actively mix and vortices appear. Turbulent flow then occurs. The transition between laminar and turbulent flow is determined by a dimensionless parameter, which is the Reynolds number Re.

$$Re = \frac{v_0 H \rho_{water}}{\eta_{water}}. \tag{5.4}$$

When Reynolds numbers are low, flows are laminar, but when they are high, flows are turbulent.

The specific value of Re_0 at which laminar viscosity is equal to turbulent viscosity depends on the geometric system and usually lies within the range of 10–25.

Thus, the internal viscosity of a liquid is:

$$\eta = \eta_{water} \text{ when } Re < Re_0; \tag{5.5}$$

$$\eta = \eta_{water} \frac{Re}{Re_0} \text{ when } Re > Re_0. \tag{5.6}$$

We have:

$$v_0 == \frac{\sigma Re_0}{\rho_{water} v_0}. \tag{5.7}$$

From this we learn that:

$$v_0 = \sqrt{\frac{\sigma Re_0}{\rho_{water}}} \approx 0.01\,\text{V}. \tag{5.8}$$

We find that the speed of the current near the surface is proportional to the wind speed and when $H = 1\,\text{m}\,(3.28\,\text{ft})$ and $V = 1\,\text{m}\,(3.28\,\text{ft})/\text{s}$, the speed of the current near the surface is $v_0 = 1\,\text{cm}\,(0.39\,\text{in})/\text{s}$, which is $Re \approx 10^4$.

A characteristic representation of this phenomenon is similar to that of a spiral twisting into the ocean's depths (Fig. 5.4).

This is, of course, a simplified model and in its pure form the Ekman spiral has only been observed under polar ice. However, understanding ocean currents is in fact much more difficult. The currents in the World Ocean are strongly influenced by surface waves, which ruin the image of a spiral going into the depths of the ocean. In addition, it is easy to understand that currents form differently near the Equator where the Coriolis force disappears.

When we were analyzing wind, we made reference to the trade winds that consistently blow in the direction of the Equator. In the vast expanse of the ocean, they stir up analogous currents, which are called *trade currents* and flow westward. In the latitudes where trade winds are found, they blow from the northeast to the southwest in the Northern Hemisphere and from the southeast to the northwest in the Southern Hemisphere. Thus, the deviation

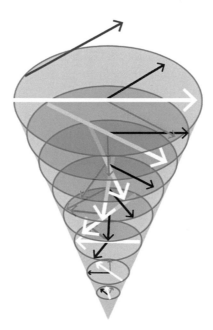

Fig. 5.4 Ekman spiral. Directions are denoted with colors: wind is dark blue, the friction force is red, the direction of the resultant is white and the Coriolis effect is orange

of the direction of currents from that of the direction of wind is indeed close to 45°, which is in line with Ekman's theory.

The Equatorial Counter Current is an eastward flowing current that is located close to the Equator between the North Equatorial Current and the South Equatorial Current. Deep countercurrents that emerge on the ocean's surface cause this transfer of ocean water to occur in the equatorial region where there are no strong winds and the Coriolis force is not at work. In other words, trade winds swirl the water into the shape of a vertically standing "donut" and the water that has risen to the top in it flows out at the Equator. This is the reason why the water temperature in the equatorial region is several degrees lower than it is in the neighboring regions of the ocean that are close to the tropics. At first, this seems strange because the Sun shines the brightest at the Equator, but the water is colder there.

Parameters of a Current Opposite to the Wind The mathematical model showing how the Coriolis force influences the direction of currents is as follows: if the wind blows along the y-axis, then the speed of a current **v** at the depth z has the following components:

$$v_x = \pm v_0 e^{-kz} \cos\left(\frac{\pi}{4} \mp kz\right);\tag{5.9}$$

$$v_y = v_0 e^{-kz} \sin\left(\frac{\pi}{4} \mp kz\right).\tag{5.10}$$

The upper signs refer to the Northern Hemisphere and the lower ones refer to the Southern Hemisphere. Currents on the ocean's surface flow at a 45° angle in the direction that the wind blows, deviating to the right in the Northern Hemisphere and to the left in the Southern Hemisphere. The velocity vector turns with increased depth, and at the depth $z = \frac{3\pi}{4k}$, it becomes channeled in an upward direction. The parameter k determines the force of the current's velocity attenuation with depth. When there is a change in geographical latitude φ, this parameter changes and is equal to:

$$k = \frac{\omega_3 |\sin \varphi|}{v_0}.\tag{5.11}$$

When this type of wind speed occurs, ocean turbulence is characterized by the Reynolds number $Re \approx 10^7$. In the ocean's subsurface layer, which is 100 m (328 ft) thick, water is most actively mixed and transferred by currents.

When currents come in contact with the shore, many interesting things happen. Continents that are located in the path of currents exacerbate this situation most of all.

Water layers near the ocean and sea coasts rise and fall when the wind direction is tangential to it. These phenomena are called *upwelling* and

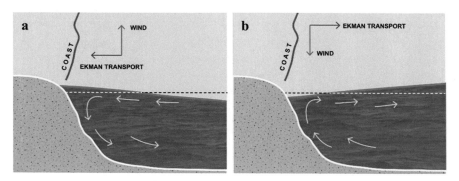

Fig. 5.5 A diagram showing the direction in which ocean currents flow in the coastal zone of the Northern Hemisphere: **a** when the water level rises (upwelling); **b** when the water level drops (downwelling)

downwelling, respectively, and a diagram illustrating how they occur in the Northern Hemisphere is shown in Fig. 5.5.

During downwelling, warm water near the coastline turns downward and goes deep into the water (Fig. 5.5a). Conversely, during upwelling deep water rises toward the surface, and when it is near the coastline, it flows in the opposite direction (Fig. 5.5b).

Upwelling and downwelling occur as a result of the Coriolis force, which causes liquid to flow perpendicular to the wind in accordance with the Ekman spiral. Coastal upwelling forces the remains of dead plankton out of the water. Nitrogen and phosphorus, as well as carbon dioxide, are actively consumed by phytoplankton, which, in turn, are eaten by fish and then eaten by other fish. It stands to reason, therefore, that coastal upwelling areas are the world's primary fishing areas.

When trade currents are close to continents or, more precisely, close to the continental shelf, they go around them and change direction. This is precisely how powerful currents form in the middle latitudes. In the Atlantic Ocean, the Gulf Stream and Brazil Current are the strongest currents; in the Indian Ocean, the Madagascar Current is the most powerful; and in the Pacific Ocean, the East Australian Current and the Kuroshio Current are the strongest. The main driving force behind these currents is not wind, but rather the water pressure generated by the continents. For this reason, these currents do not have deep countercurrents, and a rush of water that can reach a depth of up to 2 km (1.24 mi) carries a large body of water in the direction of each current. In general, the insight we have gained into the major ocean currents corresponds well with the information we have about wind and the location of the continents.

The entire biosphere—including human beings—has adapted itself to the conditions that sea currents have created, and thanks to them, numerous marine animals are able to migrate for thousands of kilometers (miles). Penguins, for example, swim in the cold Peruvian current in which they feel quite at home in order to reach the coast of Chile. The seaport in Murmansk (Russia) was built because the sea there almost never freezes because of the Gulf Stream. Thus, ocean currents are a crucial factor in keeping life on the Earth from disappearing.

5.3   Types of Waves

Most of us see the water's surface from above the division between water and air, and it is very rarely calm. The sea is associated with waves that roll in from vast expanses of open water and then crash against the shore.

Let's ask ourselves these questions: What kinds of waves are there? Why are some big and others small? Why are there sometimes waves when there isn't any wind? How do they appear in the first place? An entire section of oceanology is devoted to the study of waves. We can answer some of these questions.

Waves and their undulating movement in the oceans are characterized by an extremely wide range of wavelengths, which is the distance from the crest to the crest, and by periods, which is the interval of time it takes for two successive crests to pass a fixed point. Capillary waves are the smallest waves and have a wavelength of several inches and a period of less than a second. Tidal waves are the largest waves. The distance between their crests reaches half of the Earth's circumference, that is, about 20,000 km (12,427 mi), but their period is not the longest. Slow internal waves have the longest period due to variations in water density at different depths. Consequently, it takes months for these waves to cross the ocean.

Waves also differ by propagation speed. A gravity wave turns into a capillary wave at a wavelength of about 2 cm (0.79 in), while its speed is $v \approx 20\,\mathrm{cm}\,(7.87\,\mathrm{in})/\mathrm{s}$. A stone thrown in the water causes gravity waves to appear, which have a speed of approximately $v \approx 30\,\mathrm{cm}\,(11.81\,\mathrm{in})/\mathrm{s}$. Long gravity waves travel in the ocean at speeds comparable to that of a modern-day airplane. In an ocean with an average depth of $H \approx 1\,\mathrm{km}\,(0.62\,\mathrm{mi})$, the speed of an ocean wave is $v \approx 360\,\mathrm{km}\,(223.69\,\mathrm{mi})/\mathrm{h}$. However, in an ocean with a maximum depth of $H \approx 10\,\mathrm{km}\,(6.21\,\mathrm{mi})$, the speed of an ocean wave reaches $v \approx 1000\,\mathrm{km}\,(621.37\,\mathrm{mi})/\mathrm{h}$.

Wide ranges of wavelengths and periods of undulating wave movement in the ocean can be clearly illustrated by using a spectral graph as shown in Fig. 5.6. Various types of waves, which are classified according to wave period, are plotted along the horizontal axis of this graph. The vertical scale shows the total amount of energy that is found in waves of a certain length at any time and in all of the oceans combined.

The same diagram shows the factors that create or obstruct undulating wave movement, as well as those elements that impact upon its propagation. For example, since the water on the crest of a wind wave is above average sea level at that location, wind works against the force of gravity in order to raise it to that level. In this type of situation, wind acts as a disturbing force. The Moon's gravitational pull generates tidal forces. Tsunamis occur due to seismic activity on the sea floor, which most often comes from earthquakes. In the case of wave movement, a disturbing force also works against the force of gravity. Therefore, it is specifically gravity that plays a fundamental role in

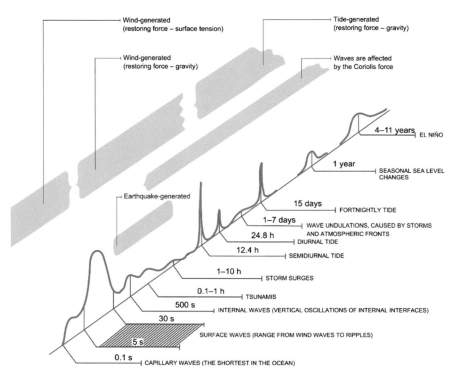

Fig. 5.6 Physical characteristics of different types of ocean waves. The numbers indicate the wave period; the values along the vertical axis show the amount of energy of each wave type

making the surface of the water become calm again. However, when capillary waves form, the wind overcomes the surface tension force on the water's surface. Thus, in this case surface tension is the primary restoring force. The Coriolis force complicates the situation regarding long-period waves, as it causes changes in the direction of wave propagation.

Let's consider some different types of waves in greater detail.

> Capillary waves are the shortest type of waves observed on the water's surface and are stirred up by wind as a result of friction between two fluids—air and water.

If we take into account the small distance between crests and the fact that capillary waves move very quickly, it becomes clear that the period of capillary waves is the shortest of all wave types. As the wind starts to blow, these waves form and cause the first changes that we see on the water's surface.

While standing in the early morning hours on a steep river bank, we can feel how the calm air gives way to a light breeze and patches of small ripples, which are sometimes called *cat's paws*, appear and disappear on the water's surface. This is exactly where capillary waves, which have a wavelength of only 2–5 cm (0.79–1.97 in), develop. When the wind blows, it causes air friction, which makes ripples on the water surface that form small waves. But the surface tension force on the water keeps trying to make it smooth again, which requires that a minimal amount of energy is exerted. In this way, capillary waves lose their kinetic energy, which is directly converted into thermal energy due to the molecular viscosity of water. Capillary waves are often indistinguishable, as they are overshadowed by other waves that are hundreds of meters (feet) long. Nevertheless, they always appear just as soon as the wind speed exceeds several meters (feet) per second. What is more, since the wind is almost constantly blowing over the ocean, it is extremely unusual to see the water's surface have a mirror-like smoothness. Capillary waves are not caused solely by the friction between wind and water.

A very perceptive onlooker will notice that capillary waves suddenly appear in the water just in front of the crests of very short waves when their steepness increases and they become unstable. But instead of surging forward as large wind waves do, it is as if the crests slide forward, thus causing a series of ripples to appear. These are also capillary waves and they perform an important intermediary role in dispersing energy from large waves. Thus, these capillary waves act as a "short circuit" when large waves continuously accumulate energy.

> Gravity waves form in the field of gravity, which attempts to push a distorted fluid surface back toward its equilibrium position.

(Do not confuse gravity waves with gravitational waves, which were recently discovered and form when large accelerating bodies are in motion!)

The force of gravity, which returns the fluid to a state of equilibrium, develops as a result of the difference between the height of the crest and the trough of the wave. When considering the gravitational wave theory, two limiting cases should be underscored. The first is called the *shallow water approximation*. Such an approximation is realized when the wavelength L (i.e., the average distance between adjacent crests) is much greater than the depth of the body of water. In this case, the wave speed v is related to the depth of the body of water h by the formula:

$$v \approx \sqrt{gh}. \tag{5.12}$$

If this fails to happen, then when $L \ll h$ (i.e., the approaching deep water), the wave speed is determined by a different equation:

$$v \approx \sqrt{gL}. \tag{5.13}$$

These formulas make it possible to explain why waves bend as they move toward the shoreline. If a wave were to move toward the shoreline at an angle, then its section closer to the shoreline would begin to slow down because the water's depth there decreases. Therefore, the section of the wave that is the furthest from the shoreline surges forward and bends the leading edge of the wave. In deep water, a wave's speed depends exclusively on its wavelength.

Speed of Capillary Waves One additional physical quantity should be added to the speed of capillary waves, which is the surface tension coefficient of water σ_{water}. It has the dimension N/m. The speed of such a wave is described by the formula

$$v \approx \sqrt{\frac{\sigma_{water}}{\rho_{water} L}}. \tag{5.14}$$

This formula is suitable for all capillary waves, but when wavelengths are long, it is necessary to use the formula for gravitational surface waves. In this case, the lowest value of the speed of capillary waves is

$$v_{\min} = \left(\frac{\sigma_{\text{water}} g}{\rho_{\text{water}}} \right)^{\frac{1}{4}}. \tag{5.15}$$

In Fig. 5.7, point 2 is located at a great distance from the shoreline where the water is deeper than at point 1. Therefore, in accordance with the formula for wave speed in shallow water, the speed of the wave at point 2 will be faster than at point 1. Thus, the leading edge of wave AA' will bend (BB') and end up parallel to the shore.

Waves caused by the wind hold more energy than ocean waves of any other type. This does not mean that one wind wave in which we play on the beach has the greatest amount of energy. No, rather, it means that every second more energy is stored in the endless number of wind waves that move in the oceans' vast surface than in any other type of waves. This corresponds to the high and wide energy peak between the 5 and 30 s periods (see Fig. 5.6). Most of the energy stored in the oceans' wind waves eventually reaches the coastline and dissipates due to turbulence in the surf zone.

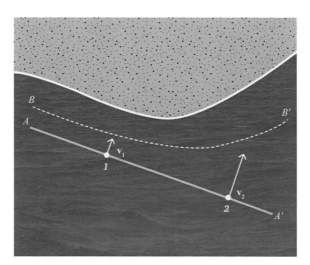

Fig. 5.7 Curvature of the water's surface in shallow water

The energy density in waves increases in proportion to its square height. The importance of this energy is colossal: the crest in a wave with a period of 10 s and a height of 2 m (6.56 ft) by 1 m (3.28 ft) in length has enough energy to light 250 incandescent light bulbs or 10 times more energy than is found in energy-efficient light bulbs.

However, this energy is unevenly distributed throughout the World Ocean. Winds are the driving force that stir up these surface waves. Therefore, as would be expected, the waves with the greatest amount of stored energy form in the zones where near-surface westerly and easterly winds blow (in the zone between 40 and 50° south of the Equator). Because of the fact that the wind blows in this area all around the Earth without dying down, the longest surface waves of all form here: the length of some of them exceeds 500 m (1640.42 ft) and their speed is 25 m (82.02 ft)/s. It should not come as a surprise that navigating around Cape Horn, that is, sailing around the tip of South America (at approximately 54° south of the Equator), was an arduous challenge for the first sailors and a test of their strength and courage, as well as the reliability of their ships.

When a storm occurs at sea, waves suddenly rise up and this, combined with strong gusts of wind, frequently causes these waves to come crashing down (Fig. 5.8). This results in the formation of white crests, or the so-called *whitecaps* on the water. Whitecaps also form when waves roll into shallow water, which happens when a wave's height a is congruent to the ocean's depth H. In this case, the speed of the top of the wave, which is equal to $\sqrt{g(H + a)}$, significantly exceeds the water's speed at the wave's base, which is equal to $\sqrt{gH}$. It is as if the top of the water overtakes its base, which is what makes the wave crash down.

The laws of physics that explain why waves crash down are quite complicated and have yet to be fully explored in depth. However, oceanographers know that energy cannot be dissipated through short waves and capillary waves fast enough to pull down the high crests of giant waves and prevent them from crashing down. At the same time, short waves and capillary waves slow down tidal waves and cause them to move upward and then crash down with tremendous force.

Waves' motion and movement have long helped sailors. The helmsman always guided the ship in the direction that waves were rolling because of which they would lap up against the ship's stern. In order to prevent them from surging over the stern, the ship's sailors hung canvas bags filled with oiled rags over the side of it. The fat in the oil reduced the amount of surface tension in the water and prevented slow short waves and capillary waves from forming and, in so doing, reduced the braking force of large waves. This

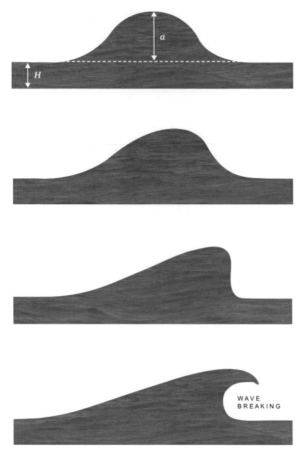

Fig. 5.8 Changes in a wave's profile as it approaches the shoreline

meant that enormous waves would pass under ships and they would not crash down. This was called *oiling troubled waters*. Such a practice would not be feasible today because ships have become much larger than they used to be and, in addition, this is harmful to the environment.

Tides are in the middle of the scale (see Fig. 5.6). Their periods range from 12 to 24 h. These fast-moving waves travel hundreds of kilometers (miles) per hour, but their width from one crest to another is so tremendous that a significant amount of time is needed in order for them to complete one cycle.

Waves with incredibly long periods are located at the far end of the spectrum (Fig. 5.6). For example, slow changes in ocean currents, which are caused by seasonal changes in wind patterns, can be considered as a wave-type disturbance and have a period of a year or even several years. This is the case with an El Niño event (i.e., currents in the Pacific Ocean that greatly affect

Fig. 5.9 A snapshot of waves in the open water

the climate of the entire planet), which is related to the so-called Southern Oscillation.

In conclusion, let's say a few words about the foam that forms on sea waves. Just what exactly is it? It is a large number of air bubbles separated by a thin film of liquid. The main reason that air bubbles form in seawater is wind waves, but bubbles can also occur in heavy rain, for example. The bubbles that are found in seawater are usually very tiny—less than 0.5 mm (0.020 in) in diameter. When air bubbles rise to the surface of the water, they burst and thus emit a spray of salt water. This explains exactly why salt particles are in the atmosphere, which, while walking along the shoreline, we refer to as "sea air."

When we look at the sea, we can simultaneously see that waves of all different types appear (Fig. 5.9). The smallest capillary waves produce tiny specks of reflected sunlight. Short gravity waves that give the water's surface a rippled appearance are the next in size after capillary waves. Still longer waves are ordinary wind waves.

5.4   Sea Swells

After the wind has forcefully blown a body of water, it calms down, but the waves that were formed keep on rolling for tens and hundreds of kilometers (miles).

When waves rise and fall but there is no wind this produces a sea swell.

These waves are quite large and can reach many tens of meters (feet) in length. As we already know, the effects that surface tension cause for such waves may be ignored. Therefore, when analyzing a sea swell, only two forces need to be taken into account—pressure and gravity.

The movement of a body of water can be thought of as the movement of water layers. We will denote the velocity of a wave as $\mathbf{c}$, and the velocity of a small volume element of water relative to the wave profile as $\mathbf{V}$ (Fig. 5.10). Then the velocity of this element in the frame of reference related to the sea floor or seashore will be equal to:

$$\mathbf{v} = \mathbf{V} + \mathbf{c}. \qquad (5.16)$$

If we were moving with a wave at the speed of its relief $\mathbf{c}$, i.e., if we moved into the corresponding frame of reference, then the water's movement would appear as a current moving along a motionless contour. If we assume that the water does not seep out of this given layer, then it is easy to understand that the curves in each of the layers become less well defined with depth (Fig. 5.11). When we are very deep in the water, we can see how water layers flow horizontally.

This is due to the fact that the thickness between the crests is greater than the thickness between the troughs for the given layer because it is thinner where the current's speed is faster. However, the speed by a wave's crest is minimal and reaches its maximum when the crest rolls down; in other words, in the area by the trough (Fig. 5.12).

Two factors—a change in pressure and gravity—cause a small volume element of water in a wave to accelerate. Gravity balances the change in pressure on the upper and lower boundaries for each specific water layer.

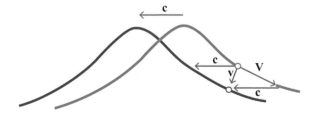

Fig. 5.10 Velocity components making up a volume element of water in a wave

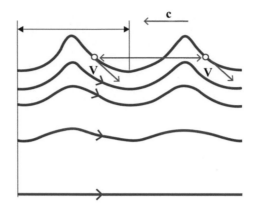

Fig. 5.11 Changes in the profile of a wave's water layer with depth

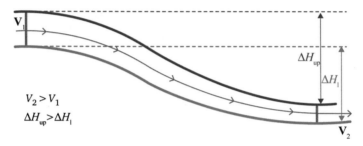

Fig. 5.12 A reduction in the thickness of the water layer from the wave's crest to its trough

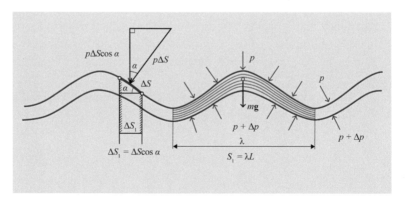

Fig. 5.13 Small sloping section of the water layer

A Pressure Drop in a Water Layer In order to determine the acceleration rate of water particles in a wave, we will first consider the pressure that acts on the upper and lower boundaries of a water layer. We will denote the pressure on the upper boundary of a specific thin layer of water as p, and the pressure on the lower one as $p + \Delta p$, the width of the layer as L, and the water's mass in this layer as m. We will analyze the small, sloping section of the layer with the area ΔS as shown in Fig. 5.13.

The force that is acting on this patch of water is equal to the product of pressure and area, while the vertical component of force is $p\Delta S\cos\alpha$ where α is the tilting angle of the wave's profile at a certain point. If we sum up all of the vertical components of force, we can find the total pressure force acting on the top boundary of the designated layer. This force is directed downwards and is equal in magnitude to $p\lambda L$. We will reason in a similar way regarding the lower boundary of the water layer. There the total pressure force is directed upwards and is equal to $(p + \Delta p)\lambda L$. The difference between these forces is balanced by the force of gravity mg, which acts on the water layer. From here we find:

$$\Delta p = \frac{mg}{\lambda L}. \tag{5.17}$$

Since the bend in the layer does not change the water's mass in it, the drop in pressure remains the same as it would have been in calm water.

Characteristics of Particle Movement in a Wave We will focus on a patch of a thin layer of water as shown in Fig. 5.14.

A body of water that passes through this given area is completely renewed during $T = \frac{\lambda}{c}$, and in a unit of time, a body of water equal to $\frac{m}{T} = \frac{mc}{\lambda}$ passes through it. In the course of time Δt, a mass $\Delta m = \frac{mc}{\lambda}\Delta t$ passes through this area.

The force of gravity that acts on this area is $g\Delta m$ and is channeled perpendicularly to the direction of the wave's movement. On the other hand, as we already know, the force that is

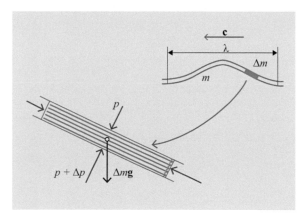

Fig. 5.14 Forces acting on the thin water layer

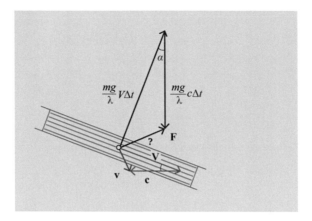

Fig. 5.15 Total force acting on the thin water layer

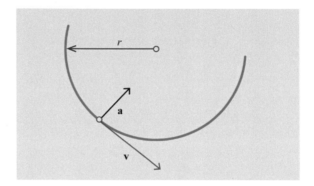

Fig. 5.16 Trajectory of the volume element of the liquid

caused by a pressure difference on the upper and lower boundaries of this area is equal to:

$$\Delta p L V \Delta t = \frac{mg}{\lambda} V \Delta t \qquad (5.18)$$

and is channeled perpendicularly to the velocity vector of current **V**. By looking at Fig. 5.15, we see that the total force results from the sum of the velocities **V** + **c** with a right-angle rotation multiplied by the constant $\frac{mg\Delta t}{\lambda}$.

Taking into account the fact that the velocity of a fluid particle in relation to the ocean floor is **v** = **V** + **c**, the total force can be written the following way:

$$F = \frac{mg}{\lambda} v \Delta t. \qquad (5.19)$$

Now we can find the rate of acceleration:

$$a = \frac{F}{\Delta m} = g\frac{v}{s}.$$ (5.20)

The acceleration **a** is channeled perpendicularly to the velocity **v**.

We know that steady acceleration, which is perpendicular to velocity, results in constant motion around a circumference. We will designate its radius as r (Fig. 5.16).

When such movement occurs, centripetal acceleration is equal to $a = \frac{v^2}{r} = v\omega$ where $\omega = \frac{v}{r}$ is the rate of rotation. Using the formula for acceleration that we previously found, we get

$$\omega = \frac{g}{c}.$$ (5.21)

An explanation of the way that fluid particles move when a sea swell forms will now be provided. Circumferences that have a constant radius in this layer of water and the same frequency are used to illustrate small volume elements of water. In each of these layers, motion occurs in a similar way and only the radius of each circumference changes. Over time the shift between the radii at the angle $\Delta\varphi = \omega\Delta t$ remains constant, while the profile shifts as an entire unit with the speed c (Fig. 5.17).

After the period of time $T = \frac{2\pi}{\omega}$ (i.e., the period of revolution), the fluid particle returns to its original position on the circumference but only in the next wave (Fig. 5.18).

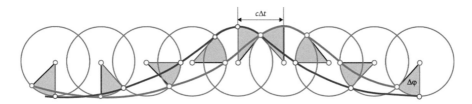

Fig. 5.17 A general look at the trajectory of a fluid particle in a wave

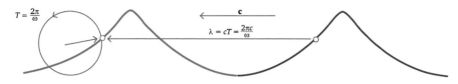

Fig. 5.18 The movement of a fluid particle in a wave within one period

Within the time T, the wave travels the distance $\lambda = cT = \frac{2\pi c}{\omega} = \frac{2\pi c^2}{g}$. We will then get a formula for the speed of a wave when deep water is getting closer:

$$c = \frac{1}{\sqrt{2\pi}}\sqrt{g\lambda}. \tag{5.22}$$

As a result, one may ascertain that the water particles in sea swells have zero average speed; in other words, they fully conform to the description of transversal waves taught in high school physics courses where they are said to oscillate only when vertically aligned. However, when we consider how these particles move, it becomes clear that their motion is not vertical at all; rather, they circumscribe the trajectory of a circumference. Additionally, the greater the water depth, the smaller the radius of the circumference that they circumscribe. What is more, when these particles are deep in the water, they rapidly lose force and at a depth equal to several wavelengths, they are practically motionless. This is exactly why waves never pose a problem for submarines that move under water; the same, however, cannot be said for surface vessels.

Wave Attenuation with Depth We will now consider two volume elements of a liquid that are located on the top and bottom boundaries of a moving water layer. The motion trajectory of these elements is a superimposition of translational and rotational motion, but the centers of the circumferences will be at different depths even though the circumferences themselves will have different radii (Fig. 5.19).

We will make use of the fact that liquid is constantly flowing in this layer and choose two cross sections of it. We will isolate the first one at the point between the vertices of the profile where the radius vectors of the volume elements are directed upward and the speed of motion is equal to $c - v$ and is directed horizontally.

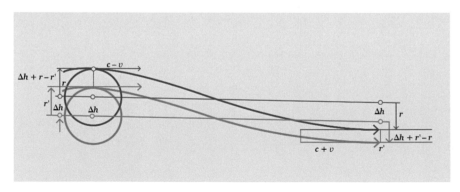

Fig. 5.19 Paths of two volume elements of the liquid on the top and bottom boundaries of the water layer

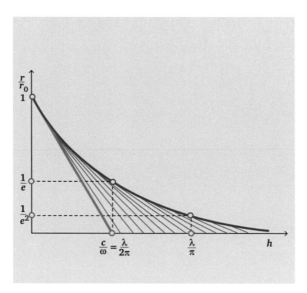

Fig. 5.20 Trajectory radius as a function of the water depth

We will isolate the second cross section at the point between the troughs of the profile where the radius vectors of the volume elements are directed downward and the velocity is equal to $c + v$. Let's assume that Δh is the distance between the centers, the radii of the upper circumference r and the lower circumference r'. The vertical size of the layer between the crests is equal to $\Delta h + r - r'$, while between the troughs it is equal to $\Delta h + r' - r$. The flow of liquid through the two cross sections we chose is equal to:

$$\left(\Delta h + r - r'\right)(c - v) = \left(\Delta h + r' - r\right)(c + v). \qquad (5.23)$$

From here it follows that:

$$\Delta r = r' - r = -\frac{v}{c}\Delta h. \qquad (5.24)$$

As one would expect, the radius of the circumference decreases with depth or, in other words, when the wave dies down. Since $v = \omega r$, a change in the radius will be proportional to its value:

$$\frac{\Delta r}{r} = -\frac{\omega}{c}\Delta h. \qquad (5.25)$$

After starting with a circumference that has the radius $r = r_0$ on the water's surface, if we slowly move down, it is diagrammatically possible to find the value of the circle's radius at any water depth as shown in Fig. 5.20.

If in the final equation we change our focus to infinitesimal changes $\Delta r \to dr$ and $\Delta h \to dh$, we will get the differential equation:

$$\frac{dr}{r} = -\frac{\omega}{c}dh. \tag{5.26}$$

Since we understand mathematical analysis, it is easy for us to write the solution:

$$r = r_0\, e^{-\frac{\omega}{c}h} = r_0\, e^{-\frac{2\pi}{\lambda}h}. \tag{5.27}$$

According to the laws of exponents, a wave loses strength the deeper it goes. The intensity of the undulating movement of waves is much less in deep water than on the surface where it is equal to wavelength. Thus, when waves get closer to deep water, the ocean floor does not exert an influence on them.

5.5   Tsunamis

Everyone has heard of tsunamis—those gigantic waves that have so much destructive power (Fig. 5.21).

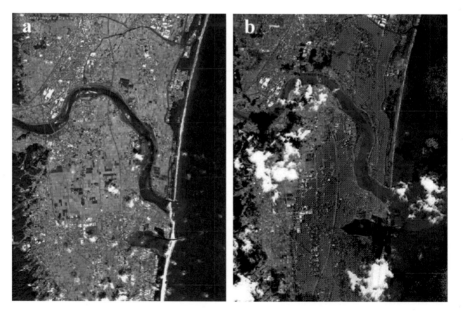

Fig. 5.21 A tsunami is shown on these pictures taken from outer space: **a** as it approaches the shore, **b** its after-effects

Tsunamis are long and high waves that occur as a result of some type of strong impact against the ocean's entire water column.

Tsunamis are primarily caused by underwater earthquakes during which a plate under the ocean floor suddenly rises or falls. Any underwater earthquake can cause a tsunami, but strong tsunamis occur because of strong earthquakes (with a magnitude higher than 7.0). As a result of each underwater earthquake, a series of successive tsunami waves are propagated. More than 80% of tsunamis occur in the coastal zones of the Pacific Ocean, which are areas of high seismic activity. The Ring of Fire is shown in Fig. 5.22.

In the open ocean at an average depth of 4 km (2.49 mi), tsunami waves propagate at extremely high speeds—up to 200 m (656.17 ft)/s (or 720 km [447.39 mi]/h). Seismic sea waves or tsunamis usually occur due to the movement of plates under the ocean floor, which is tens and even hundreds of kilometers (miles) long. Hence, the wavelength of tsunamis (i.e., the distance between the crests) in the open ocean reaches hundreds of kilometers (miles),

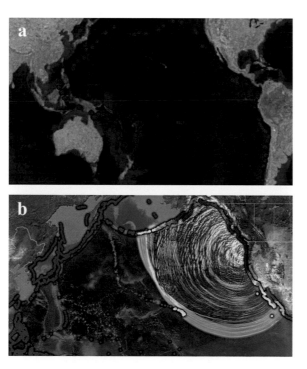

Fig. 5.22 Tsunami model animation: **a** the Ring of Fire in the waters of the Pacific Ocean in the twentieth and beginning of the twenty-first centuries; **b** a tsunami wave spreading out in the Pacific Ocean as a result of an underwater earthquake

while the height of its waves rarely exceeds 1 m (3.28 ft). Waves such as these do not pose a threat to ships. Indeed, a passenger on the deck of a ship that is sailing in the open ocean would not even feel it gradually rise up 1 m (3.28 ft) and then dip back down again.

However, tsunamis become quite dangerous near the coast when they roll in to shallow water. The reason for this is because near the coastline waves' speed and wavelength sharply decreases, while their height, on the other hand, increases. The ocean floor slows down waves' lower layers, while the upper layers surge toward the lower ones (Fig. 5.23). Furthermore, their kinetic energy turns into potential energy.

We can calculate the height of a tsunami wave when it reaches shallow water with this formula:

$$H_{\text{shallow}} \approx H_{\text{deep}} \left(\frac{h_{\text{deep}}}{h_{\text{shallow}}} \right)^{\frac{1}{4}}, \qquad (5.28)$$

where H_{deep} is the height of the initial wave in deep water, h_{deep} is the depth of the ocean in deep water and h_{shallow} is the depth of the ocean near the coast in shallow water.

If, for example, a tsunami wave formed in the ocean with a depth of 6 km (3.73 mi) and a height of 1.5 m (4.92 ft), while approaching the shore at a depth of 10 m (32.81 ft) its height would increase to 7.5 m (24.61 ft). Tsunamis become even more dangerous when they enter narrow straits and bays because then the height of their waves increases by the factor $\sqrt{\frac{\Delta_0}{\Delta}}$ where Δ_0 is the width of the area of water at the entrance to the bay and Δ is the

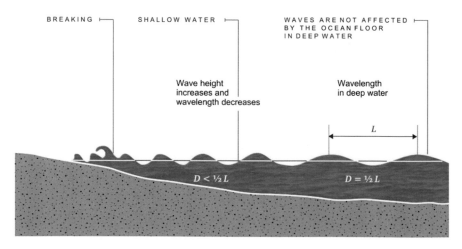

Fig. 5.23 Ocean waves' behavior as they approach the coast (D stands for "depth")

width at a specific point. Tsunami waves up to 30 m (98.43 ft) and higher
have been recorded in straits.

Height of Tsunamis by the Coast We will estimate the height of a tsunami wave based on
the law of conservation of energy. The total wave energy is the sum of kinetic and potential
energy. The kinetic energy of a small volume element of water moving along a circular path
with the radius r is:

$$E_x = \frac{mv^2}{2} = \frac{m\omega^2 r^2}{2} = \frac{\pi mgr^2}{L}. \tag{5.29}$$

Here we have used the formula for the upstroke velocity of a fluid particle that is in a
wave and is traveling along the circular path $\omega = \sqrt{2\pi}\sqrt{\frac{g}{L}}$. Since the potential energy of a
fluid particle changes while it is in motion, it therefore makes sense to talk about the potential
energy that is averaged within one period of motion. This value is determined by the degree
to which the centers of the moving water particles exceed their locus in the case that there is
no undulating wave movement (Fig. 5.24).

One can prove that the average level of particles on the wave's surface exceeds the level of
calm water by the amount:

$$\Delta d = \frac{\pi r^2}{L}. \tag{5.30}$$

It follows that the potential energy of a fluid particle found in a wave, which is averaged
according to the period of motion, $E_p = mg\Delta d$, is exactly equal to its kinetic energy. In
order to calculate the total energy of the entire wave, we must consider the formula that we
obtained earlier for the attenuation of the wave's intensity with depth, which is:

$$r = r_0 e^{\frac{2\pi}{L}h}, \tag{5.31}$$

and calculate the total amount of energy of the fairly thin layers of water. Such a sum total
will include an infinite number of components (this operation in mathematics is called an

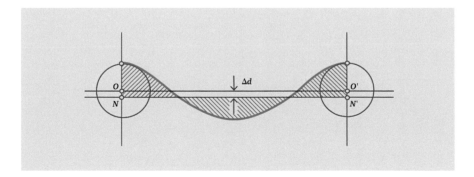

Fig. 5.24 In calculation of the potential energy of the wave movement

integration). Thus, the total energy of the tsunami will be equal to:

$$E = 2\pi \rho_{\text{water}} g \frac{r_0^2}{L} \int\limits_0^\infty e^{-\frac{4\pi}{L} h} dh = \frac{1}{8} \rho_{\text{water}} g H^2. \qquad (5.32)$$

This value corresponds to the energy found in 1 m (3.28 ft) of the wave's leading edge and has the dimension J/m. The total energy of a tsunami with the wavelength L will be equal to:

$$E = \frac{1}{8} \rho_{\text{water}} g H^2 L, \qquad (5.33)$$

where H is the height of the wave's crest.

When a wave washes into shallow water, its energy does not change; therefore, $H_{\text{deep}}^2 L_{\text{deep}} = H_{\text{shallow}}^2 L_{\text{shallow}}$. Since wavelength is related to a wave's speed and one can use the approaching shallow water in order to calculate a tsunami's speed, $L = vT = \sqrt{gh}T$ (T is the wave period and h is the depth of the body of water), we get the formula:

$$H_{\text{shallow}} \approx H_{\text{deep}} \left(\frac{h_{\text{deep}}}{h_{\text{shallow}}} \right)^{\frac{1}{4}}. \qquad (5.34)$$

Tsunamis are very different from ordinary wind waves. Since tsunamis in the open ocean usually reach a height of about 1 m (3.28 ft), they are practically unnoticeable in contrast to "ordinary" ocean waves that are tens of kilometers (miles) in length. Moreover, if wind waves cause only the surface layer of water to oscillate, during a tsunami cyclical motion occurs along the ocean's entire water column right down to the ocean floor (Fig. 5.25). This is precisely why the topography at the bottom of the ocean affects the propagation of tsunamis because they "stumble" on seamounts, which can be located deep within the ocean.

The speed at which a tsunami travels is determined by wave movement because of water that overflows into the field of gravity. Thus, the acceleration of gravity g and the depth of the ocean at a given location h influence a tsunami's speed:

$$v = \sqrt{gh}. \qquad (5.35)$$

Wavelength does not enter into this formula at least as long as it is much more than the water's depth. If we substitute numbers and use a depth of 4 km (2.49 mi), this will actually result in 200 m (656.17 ft)/s! However, as

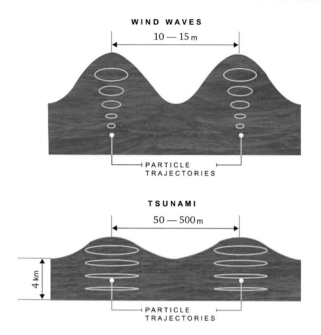

Fig. 5.25 Particle trajectory of water in wind waves and tsunamis

soon as the wave rolls into shallow water, its speed suddenly falls: at a depth of 10 m (32.81 ft), its speed is only 10 m (32.81 ft)/s (Fig. 5.26).

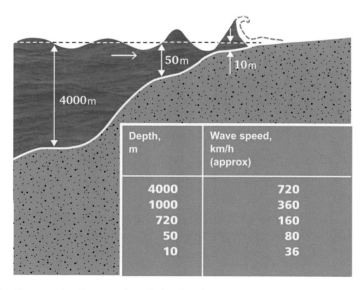

Depth, m	Wave speed, km/h (approx)
4000	720
1000	360
720	160
50	80
10	36

Fig. 5.26 Changes in the speed and depth of a wave as it approaches the coast

It is important to note that wave speed and current speed should not be confused. Water itself moves quite slowly: its speed is about a/h times less than wave speed where a is wave amplitude. This means that in the open ocean current speed is about four times less than wave speed.

Two conclusions can be drawn from this formula about the speed of tsunamis. First, when waves roll into shallow water, their height increases. From a physical standpoint, this can be understood the following way. Since the wave front suddenly slows down when it gets close to the shoreline, its subsequent sections then catch up to it. This results in the upward surge of a wave. Second, as waves move closer to the shoreline, they tend to break. Since h is the depth that differs in the crest and the trough, the speed of the crest is greater. When the wave approaches the shoreline, it breaks and spills forward. The shallower the body of water, the stronger the impact of the break.

In shallow water near the shoreline, tsunamis can ultimately reach a height of tens of meters (hundreds of feet); what is more, the crest crashes down from such a tremendous height that it causes destruction. The highest waves, which stand 50–70 m (164.04–229.66 ft), can occur in narrow, V-shaped bays. However, the area along the coast, where the entrance to bays is relatively narrow, poses less of a threat. Since tsunamis usually generate a series of waves, the interval between their wave period can be very long—more than one hour, for example—due to their extremely long waves. People who live in potentially dangerous areas know that after a tsunami wave has retreated, it is important to wait several hours before returning to shore.

Lastly, we will estimate the amount of energy in tsunami waves. During an earthquake, the initial displacement of the ocean occurs above the focus. We can assume that at that moment all of the tsunami's energy exists in the form of potential energy to push up the liquid column over the focus. We will denote the average height of the displacement of the ocean as d. Then the potential energy will be expressed by the formula $E = \frac{1}{2}\rho_{water}gd^2S$, where S is the area of the focus. We will use the focus size 100×1000 km ($62.14 - 621.37$ mi), which is typical for powerful earthquakes. For a focus with an average height of displacement $d = 0.5$ m (1.64 ft), this turns out to be approximately $E = 10^{14}$ J, which is equal to the amount of energy of the bomb dropped on Hiroshima. However, according to oceanographers' calculations, a tsunami that struck in the Indian Ocean on December 26, 2004 had 390 times more energy! This means that the average height of the initial perturbation level was about 10 m (32.81 ft).

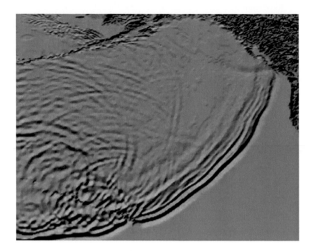

Fig. 5.27 A computer model of the interference of tsunami waves

Just like any wave, a tsunami may cause wave interference (Fig. 5.27). If a wave suddenly arrives at a given point by having traveled along several different paths (i.e., because of refraction and reflection), then it is superimposed on itself. This means that locally there will be a decrease or increase in the resulting wave's amplitude. Patterns caused by wave interference because of an increase in amplitude are usually very complex and significantly depend on the profile of the ocean floor. Since changes in depth greatly impact upon waves, even relatively small (i.e., several hundred meters [feet] high) but steep seamounts or cracks can affect wave interference and subsequent propagation. Insufficient knowledge about the profile of the ocean floor makes it quite difficult to predict when a wave will appear at one place or another and what its height will be. Ships have mapped certain areas of the ocean extremely well, but places still remain that they have yet to explore.

It is extremely important to learn how to accurately calculate the behavior of a tsunami that has been triggered by an earthquake in order to evacuate people from dangerous places and avoid casualties.

5.6 The Color of the Sea

In almost all of the paintings done by the world-renowned artist Ivan Aivazovsky, the main focus is waves (Fig. 5.28). Neither his predecessors nor his successors have been able to depict waves on canvas as realistically as he did. Despite the fact that Aivazovsky created all of his paintings from memory, they so accurately show every detail of lighting and the color of the waves

Fig. 5.28 Ivan Aivazovsky's painting "The Black Sea"

that they seem realistic as if the water is actually going to spill out of the picture frame onto the floor.

One cannot help but wonder why the sea has so many different colors and shades because, after all, doesn't sea water poured into a glass seem almost colorless and transparent?

The answer to this question lies in the fact that water acquires its light blue hue we know so well when the thickness of the water layer increases. The reason for this is that water absorbs rays of light in the red spectral region with somewhat greater intensity than it does rays in the dark blue spectral region. In addition, dissolved elements or suspended solids may give water various shades of color.

The structure and properties of water molecules determine the pure light blue hue in a water layer. The molecules of different substances have characteristic oscillation frequencies, which result from their structure and composition.

There are three such frequencies for a water molecule, and they correspond to the following wavelengths of electromagnetic radiation: 2663, 2740 and 6270 nm. These wavelengths correspond to the infrared spectral region and are not visible to the human eye. The maximum absorption of radiation corresponds to three characteristic wavelengths. However, light rays with similar frequencies, such as those that make up the red segment of the visible spectrum, are also absorbed. Therefore, the rays of light specifically in the red segment of the visible light spectrum are absorbed most actively and their energy is dissipated by oscillating molecules in the form of heat.

After the sea has absorbed red and yellow rays of sunlight, the reflected light that hits our eyes is blue and green (Fig. 5.29). This means that we see the color aquamarine. That being said, the more living organisms such as phytoplankton and small algae there are in the water, the greener it looks. In fact, red light is scattered less by small particles than dark blue light. The greener the sea water, the more organisms live there. However, if there are only a few phytoplankton in the sea, clear water becomes bright blue or, in other words, ultramarine. This is both a color hue and an ink color coined by Italian artists, which literally means "above and beyond the seas." On a clear sunny day, a calm sea looks turquoise.

We already know that seawater that is ultramarine or turquoise contains very few living organisms. But it is wonderful to look at an ultramarine- or turquoise-colored sea while sitting on a white sand beach or on the shore of an island made up of white limestone. There the light that penetrates the sea floor reflects the Sun's rays and a thin layer of water is also illuminated from underneath the surface.

The color of the sea obviously depends on the weather at a specific moment in time. When the sky is blue and cloudless, it makes the water look darker, but when it is overcast, the water takes on a threatening lead gray color.

Now imagine that we are studying the color of the water deep beneath its surface. In this instance, the color that we see now is quite different from what we perceive when we look at the water from above. In the first case, the color of the water absorbs sunlight, while in the second case, the color is determined by the reflection of light and depends on the color of the sky.

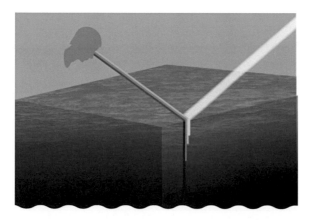

Fig. 5.29 The color of the sea is determined by special characteristics with regard to how water absorbs sunlight

The deeper one goes underwater, the less intense the color of the water is, which underwater diving enthusiasts know very well. At a depth of 25 m (82.02 ft), the color of the water becomes dark bluish-green. This is because the red and yellow rays of the sun, which are absorbed by the water column, are almost unable to reach this depth. For this reason, the scales of deep-sea fish are dark blue and violet, which makes them practically invisible just like everything else found deep underwater that has a dark color. On the other hand, the equipment used by deep-sea divers and scuba divers is always bright yellow, which makes them visible to fellow divers (Fig. 5.30).

We are diving still deeper. At a depth of 40–50 m (131.23–164.04 ft) darkness prevails almost like at night. Everything here looks dark violet. This is because the absorption coefficient in visible light is minimal for violet rays.

We have figured out what determines the color of seawater. But what accounts for the color of whitecaps that form on waves when they rise and fall at a certain strength or, to be more exact, at a windspeed of about 5 m (16.40 ft)/s (Fig. 5.31)?

Whitecaps are made up of air bubbles that randomly emerge and disappear on the top of breaking waves. Rays of falling light are repeatedly reflected from the outer edges of these bubbles, although the intensity of reflection does not depend on wavelength. In other words, all segments of the visible light spectrum reflect light the same way.

In this type of environment, sunlight is fully reflected and the very same spectral distribution of solar radiation that fell on the waves is what hits our eyes. This is the reason why sea foam appears white to us. It is also the reason why clouds and steam from a boiling tea kettle look white.

Thus, the color of the sea depends on many factors: the presence of plankton and inorganic suspended solids, the wind, the sun, clouds and visual angle. This is why the color of the sea, just like the image in a kaleidoscope,

Fig. 5.30 A deep-sea diver

Fig. 5.31 Whitecaps on waves

never repeats itself and only Aivazovsky could capture this fleeting and unique scene and transfer it onto a canvas. When looking at his paintings, one can even learn about the physical characteristics of the optical phenomena that are portrayed.

For example, Aivazovsky loved to paint scenes of the sea at night and make a moonglade the focal point of a seascape (Fig. 5.32). Let's explore the phenomenon of moonlight that is reflected off the water.

Moonglades We will now consider a scenario in which an observer looks at the water's surface on a moonlit night from a cliff at the height h (Fig. 5.33).

Fig. 5.32 "Sailing off the coast of the Crimea in the moonlit night" by Ivan Aivazovsky

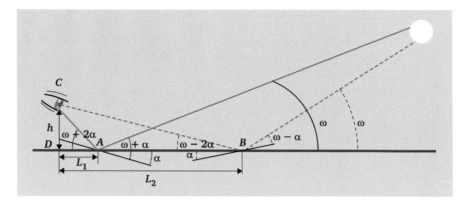

Fig. 5.33 Geometry of the moonglade formation

We will denote the angle between the direction from the observer to the Moon and to the water's surface as ω. We will assume that this angle is roughly the same for the points on the water's surface at the beginning (point A) and at the end (point B) of the Moon's path. This is justified since the distance from the observer to the Moon is much greater than the length of the moonglade, which does not usually exceed several kilometers (miles). We will speculate that there are many small waves on each patch of the water's surface that reflect the falling light and denote the point where the observer's gaze falls as C.

Let's assume that a wave's slope angle at point A has the maximum possible value α. If the wave moves in the direction of the observer, then the solar incidence angle at point A will be $\omega + \alpha$. The angle of reflection will be the same as this. However, with regard to the plane of water, the reflected ray of light will be the angle $\omega + 2\alpha$ as is shown in the illustration. If point A is the closest point to the observer and the spot from where the reflected light ray is visible to the eye, the distance from the base of the cliff to point A will be equal to:

$$L_1 = h \operatorname{ctg}(\omega + 2\alpha). \qquad (5.36)$$

The points on the water's surface that reflect the moonbeams seen by the observer will be to the right of point A for all waves with a slope less than α. The furthest point from the observer that satisfies this requirement (point B) will correspond to the slope of the wave descending at the angle α. When this occurs, the mirror image created by the wave will be tilted downward at the same angle α. The moonbeam's angle of incidence onto point B will generate the value $\omega - \alpha$ as seen in the illustration. The ray that is reflected at the same angle off the wave will generate the angle $\omega - 2\alpha$ with the plane of water.

If this ray is coming from the point furthest from the observer, then

$$L_2 = h \operatorname{ctg}(\omega - 2\alpha). \qquad (5.37)$$

Thus, the length of the moonglade will be equal to:

$$L = L_2 - L_1 = h\big(\operatorname{ctg}(\omega + 2\alpha) - \operatorname{ctg}(\omega - 2\alpha)\big)$$

$$= h \frac{\sin 4\alpha}{(\sin \omega)^2 - (\sin 2\alpha)^2}. \qquad (5.38)$$

Let's analyze this formula. When the sea is calm, the slope of the wave is zero and we have $L = 0$. In the case that $\alpha \to \frac{\omega}{2}$, the length of the moonglade approaches infinity. If the Moon is high enough so that $\omega > 2\alpha$, then a moonglade cannot appear. However, it obviously cannot be endlessly long either. The maximum length of a moonglade is the distance to the horizon.

We see that along a straight line, which is called a *track* and connects the Moon and an observer, a practically solid light glistens. The track separates into individual bright spots of light slightly off to the left and the right and then becomes blurry in the distance.

Is this really how a moonglade looks? What is the scientific explanation for it?

A moonglade occurs when a large number of waves reflect light. Each wave gives off its own separate reflection and together they form a track. Often light sources that reflect off the water's surface appear as elongated patches, that is, long tracks that run from the light source to our eyes.

Let's learn about the circumstances that cause moonglades to appear. First and foremost, there should be at least a little bit of wind or the water will reflect the moonlight like a mirror (Fig. 5.34). In addition, the water should not be too rough (in Aivazovsky's painting, it is clear that a light breeze is blowing). When the wind is strong, this causes whitecaps to form and the reflection of the Moon is very burry. Observations have shown that moonglades form at wind speeds that range from 2 to 15 m (6.56–49.21 ft)/s.

In such conditions as these, the undulating motion of waves is characterized by ripples—a large number of small waves—that randomly appear on the water's surface. The wave's slope angle, however, does not exceed the specific value $\alpha < 20 - 30°$ since, as we already know, at steeper angles waves break when whitecaps form.

In order to understand what the length of a moonglade can be, we will estimate the distance to the horizon. To do this, we will first analyze a right triangle formed by the observer's gaze and the center of the Earth (for simplicity's sake, let's assume it is shaped like a ball) and tangent to the Earth's surface (Fig. 5.35).

Fig. 5.34 The Moon's reflection: **a** when the sea is like a mirror; **b** when there is slight wave activity

Fig. 5.35 An estimation of the distance to the horizon

If h is the distance from the Earth's surface to the eye of the observer, then the distance to the horizon will be:

$$D = \sqrt{(R_E + h)^2 - R_E^2} = \sqrt{2R_Eh + h^2} \approx \sqrt{2R_Eh}. \qquad (5.39)$$

We utilized the condition that $R_E \gg h$. If the observer is at sea level, the distance to the horizon will be approximately 4.5 km (2.80 mi). If the observer looks at a moonglade from a 20-m-high (65.62 ft) cliff, the length of the moonglade increases to 16.5 m (54.13 ft).

And what determines the width of the track of a moonglade? The critical angle of rotation at which the light is still visible to the observer's eye is equal to α (Fig. 5.36). Therefore, the bandwidth (the distance between points P and P' in Fig. 5.36) will be $2h\mathrm{tg}\alpha$.

Hence, the short axis of the moonglade will retract angle $\beta = \frac{2h\mathrm{tg}\alpha}{\sqrt{l^2+h^2}}$. The formula of the major and minor semiaxes of the moonglade is, therefore, $\beta/2\alpha$. If the angle α is small, then $\beta/2\alpha = \sin\omega$ (ω is the angle at which we look at the water). From this it follows that the width of the moonglade increases depending on two factors: wind strength and the height of the Moon above the horizon.

We calculated that the moonglade is visible at a distance of about 4 km (2.49 mi) if the observer is at sea level. Surprisingly, in the painting (see Fig. 5.32) one can tell by looking at the opposite shore that it is indeed possible to estimate the distance this way. Moreover, along the edges of the moonglade and just beyond the width of the track we calculated, we see separate bright spots of light, which emerge due to certain patches of water reflecting the moonlight. We can see this in the picture as well.

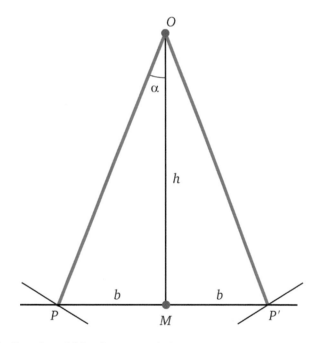

Fig. 5.36 Finding the width of a moonglade

Further Reading

1 Apel, J.R.: Principles of Ocean Physics. Academic Press (1987)
2 Byalko, A.V.: Our Planet the Earth. MIR Publisher (1983)
3 Elmore, W.C., Heald, M.A.: Physics of Waves. Dover Publications (1985)
4 Goff, J., Dudley, W.: Tsunami: The World's Greatest Waves. Oxford University Press (2021)
5 McCoy, K., Bascom, W.: Waves and Beaches: The Powerful Dynamics of Sea and Coast, 3rd edn. Patagonia (2021)
6 Pollack, G.: The Fourth Phase of Water: Beyond Solid, Liquid, and Vapor. Ebner & Sons (2014)
7 Raichlen, F.: Waves. MIT Press (2012)
8 Talley, L.D.: Descriptive Physical Oceanography: An Introduction, 6th edn. Academic Press (2011)
9 Van Dyke, M.: An Album of Fluid Motion, 14th edn. Parabolic Press, Inc. (1982)

6

Fresh Water on the Earth

Abstract This chapter tells about freshwater on the Earth's surface. First, we consider the formation of groundwater and the general features of hydrological cycle on the Earth. Then we study the physics of rivers, including their velocity and flow direction. Special attention is given to the shape of river channels, as well as meanders formation. We estimate the energy related to the flowing river. At the end, we look at waterfalls and glaciers, including the physical model of glacier flow. The physics of icebergs is also discussed.

King Solomon once remarked that rivers always empty into the sea, but it does not overflow and rivers return to their source. Only relatively recently, after comparing the amount of precipitation and the discharge of water in rivers, people began to understand why this happens. This continuous movement of water on the Earth is called the *hydrological cycle* (Fig. 6.1).

The hydrological cycle involves the transfer of water from oceans and seas into the atmosphere because of evaporation. It also includes moisture from the atmosphere that falls on the ground or into the ocean in the form of precipitation and groundwater flow into the ocean.

Each year approximately 355,000 km^3 of water evaporate from the surface of oceans and seas, primarily because of the sun. About 80% of this moisture returns to oceans in the form of rain and snow. This water is of little use to humans. The remaining 20% is needed for people to live, which means that 35 km^3 of water falls on the ground every year.

© The Author(s), under exclusive license to Springer Nature Switzerland AG 2023
D. Livanov, *The Physics of Planet Earth and Its Natural Wonders*, https://doi.org/10.1007/978-3-031-33426-9_6

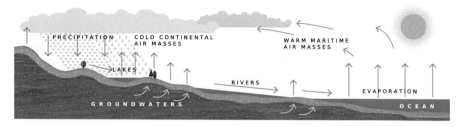

Fig. 6.1 The hydrological cycle on the Earth

In order for water to return to the ocean, it has to make a treacherous journey. When water flows high in the mountains, it may end up seeping into a glacier and then spend thousands of years moving with it before finally melting and forming riverheads in mountains. Water can also flow into lakes and stay there for a long time before ending up in a river that flows out of a lake. Furthermore, water can become groundwater and replenish underground lakes. However, if it falls as precipitation in high latitudes, water can freeze underground in permafrost regions, which, just like a refrigerator, preserve the remains of ancient mammoths.

All of this water is called *freshwater* because it does not contain any salt, which is found in the World Ocean. It is specifically freshwater that is fundamental for us to survive.

6.1 Groundwater

If you take a shovel and start digging a hole, at some depth water will almost always begin to appear. Miners and subway construction workers know this well. Water makes their job very difficult because they have to constantly pump it out of underground structures. As a matter of fact, there is a tremendous amount of water underground—almost 40 times more than the water supply in rivers and lakes.

At a certain level below the Earth's surface called the *saturated zone*, water fills most of the pores in soil and rocks. This is exactly why water will stand in a well that is dug deeper than the depth of this zone.

The water that is held in the saturated zone is called *groundwater*.

The upper limit of the saturated zone is called the *water table*. The location of the water table in relation to surface patterns is shown in Fig. 6.2. One can see that the water table is usually the same as the surface topography where it is located: it is higher beneath elevated ground than beneath valleys, while close to rivers and other bodies of water the water table is close to the Earth's surface. The unsaturated zone may even appear above the water table. This layer or so-called *lens* and others like it can be an important source of water. In India, for example, lenses accumulate water that poured down as rain thousands of years ago. This water serves as a non-renewable resource. If it is not used wisely, there will not be any more fresh water in that area and it will be impossible to live there.

Groundwater never stops moving. It trickles along small cracks in the saturated zone. This type of movement by a liquid is called *filtration*. What determines the speed of water's movement? It is clear that it depends on the permeability of rock. It is much easier for water to pass through sand than through a solid layer of mountainous rock. This speed obviously depends on differences in water pressure, which are determined by the height between the starting and ending points and the horizontal distance between them. For the most part, underground rivers flow according to the same laws as all other rivers: they proceed to low elevations, and the speed of their current depends on differences in elevation. (It is specifically for this reason that mountain rivers flow faster than lowland rivers.) The difference between them is that the filter through which water flows is extremely important for underground rivers.

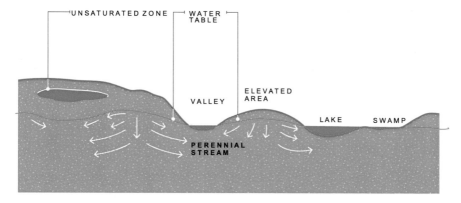

Fig. 6.2 The water table in relation to surface topography

It is possible to calculate the speed of a river's current for a specific location on the Earth's surface by using rock permeability data and the slope of the water table. The speed of the current usually ranges from a few millimeters (inches) to several meters (feet) per day. Therefore, an underground river looks completely different from what we usually imagine a river to look like. The slow movement of water is not at all like the way water flows into a channel. In order to image how an underground river flows, drop a little bit of water on a lump of sugar. The way that the water spreads over the sugar cube resembles the flow of a miniature underground river. It is important to note, however, that because of capillary force water in sugar can "flow" up as well as down. But water in an underground river always flows down because just like an ordinary river, it is propelled by gravity.

Seepage Rate of Groundwater In the mid-nineteenth century, the French scientist and engineer Henry Darcy developed a theoretical foundation for fluid filtration. Darcy's law describes the amount of fluid Q that flows through a porous filter with the length L and the area S and the difference of the fluid level above the filter and at its base ΔH, which is:

$$Q = \frac{DS}{L} \Delta H. \tag{6.1}$$

The coefficient D in Darcy's law is called *permeability*. It is related to the density p, the viscosity η of the liquid flow and the property of a porous substance, which is characterized by the hydraulic conductivity κ (it has an area dimension):

$$D = \frac{\kappa \rho g}{\eta}. \tag{6.2}$$

If we introduce a difference in fluid pressure along both sides of the filter $\Delta p = \rho g \Delta H$ and the filtration velocity, which is equal to the liquid volume flowing through a unit of the surface area in a unit of time $w = \frac{Q}{S}$, then Darcy's law can be rewritten like this:

$$w = \frac{\kappa}{\eta} \frac{\Delta p}{L}. \tag{6.3}$$

Hydraulic conductivity is very different for different types of rocks. It may reach 10^{-7} m^2 for pure gravel, it is approximately equal to 10^{-12} m^2 for gritstone, 10^{-14} m^2 for fine-grained sandstone, while it is equal to 10^{-16} m^2 for clay. Thus, water infiltrates into the soil very well through course-grained sand or gravel but much worse through fine sand or a mixture of sand and clay. However, it is practically impossible for water to pass through clay rocks.

Surface water (i.e., rivers) and underground water (i.e., groundwater) are constantly interacting. During heavy rains or periods when snow melts riverbanks overflow, river water seeps into the ground and replenishes underground reservoirs.

But as soon as a river returns to its channel, the water table becomes elevated and the river's flow is reversed. In a temperate climate this happens in summer and winter, while in a tropical one it occurs during the dry months. Groundwater provides "nourishment" to rivers in the form of streams that trickle down from steep banks or hidden springs that bubble from the bottom of a river.

In rural areas, the question of how to gain access to water is often solved by drilling a borehole into an aquifer (Fig. 6.3). A borehole pump can be put into a borehole in order to force clean water to the surface. The more water that is pumped out, the more intense the pressure difference and the more water that flows toward the borehole. If water is not pumped out of it for a long time, the borehole will fill up with sand. Then it is very possible that no water will come out of it at all because the pump will be completely covered in sand and unable to function. If the water in a well or a borehole rises above the level at which it was discovered while drilling, this is called an *artesian well*. In artesian wells, more water flows in than flows out because of a pressure difference in various parts of the aquifer. In some cases, water from an artesian well can even splash and gush out onto the ground.

6.2 Rivers

Since ancient times, river banks have been an auspicious place for people to live. Rivers provided water, fish and there were more animals and vegetation along their banks. Once boats were built, rivers also became the main waterways for travel. Thus, the Nile was the primary "federal highway" for ancient Egypt, while the Tigris and Euphrates flowed through the territory of ancient Babylon. In Russia, the Dnieper and Volga were the largest waterways.

For thousands of years, rivers have carried water thousands of kilometers (miles), cutting through mountain ranges in the process. Rivers are a symbol of eternal movement and life. Where did they originate and why doesn't their water ever dry up?

Very often rivers begin in the spot where groundwater rises to the surface. If this happens on a hillside or in a ravine, groundwater feeds streams; if this is on a low-level area, a swamp forms. The Ural River, for example, begins with springlets that flow from the slopes of mountains, while the Volga River

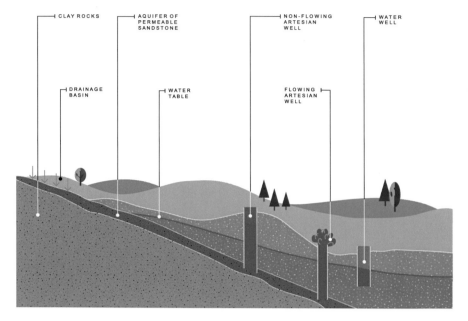

Fig. 6.3 Drawing water out of the depths of the Earth in a rural area

flows out of a swamp. The Zeravshan River in Central Asia originates from a glacier cave. Many rivers in the Caucasus and Central Asia are fed from the water in glaciers that flow down from the mountains. However, regardless of how water is supplied to rivers (rain, snow, glaciers or soil), atmospheric precipitation is the primary source from which it comes. It is worth noting that even the largest rivers on the Earth start with nothing more than a stream as their source. The exceptions to this are rivers that flow from large lakes, for example the Angara or Neva. Most often many rivers and streams flow into large lakes, but only one flows out of them. Thus, 336 rivers flow into Lake Baikal, but only the Angara River flows out of it. Only the Neva River flows out of Lake Ladoga, for example, and only the Svir River flows out of Lake Onega, although many streams and rivers flow into both of them.

This unique characteristic is due to the fact that water flows out along the deepest channel; all other channels are higher than the water level in a lake. In Fig. 6.4, a lake is shown in which the water level is lower than the four existing outlets. When the water level rises, water will flow out of the reservoir along outlet 1, which is the deepest. It will then be the river's channel (Fig. 6.4b).

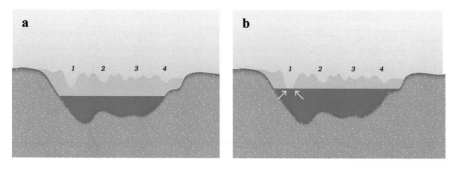

Fig. 6.4 Formation of an outlet when the water level in a lake rises

It is unlikely that two outlets exist on the same level. If this happens, it is usually in a young, deep lake. With time one of the channels will erode more than the other and all of water will flow out through only one channel. The other one will gradually dry up.

However, river bifurcation, which is when a river separates into two or more streams that never merge into one, is well known in hydrology. Sometimes these streams even flow into different basins, but this happens very rarely. One of the few examples that can perhaps be considered to this end is the Orinoco River in South America of which the Casiquiare River is a distributary. However, in a few thousand years, one of these rivers will form a deeper channel and they will merge into one. Most often seasonal bifurcation occurs when some of the water spills over a watershed and ends up in another basin because a river overflows.

Near the estuary of large rivers, their channel branches into several others and sometimes even a hundred smaller ones. This type of river branching is called a *delta* because when looking at a map, it resembles the Greek letter "Δ." River fragmentation usually results when a river that empties into a lake or sea develops a buildup of sediment and other solid material carried by currents, which results in a decrease in its velocity. Deltas form because of complex mutual effects between river runoff, tides, strong winds that cause water to pile up and so forth.

How are rivers supplied with water? Rivers, both small and the very largest, are replenished by both groundwater and surface water. We already know how groundwater feeds rivers. Tributaries, as well as spates and freshets, help to fill rivers with surface water.

A freshet is a prolonged rise of the water level in a river, which repeatedly happens at certain times of the year.

A river's primary supplier of water determines the cause and time of the freshet. For example, if a river is fed first and foremost by snow, then the freshet occurs in spring, which is when snow most actively melts. If a river receives its water from rain, the freshet occurs during the rainy season. Thus, in central Russia freshets take place in fall, but for rivers that get their water from melting glaciers, freshets take place in summer. Two freshets may occur when rivers are fed by different sources. This happens, for example, near mountain rivers in spring when snow melts and in summer when glaciers melt.

Rivers in southern areas often have a very active freshet period, and when it is finished, the water level drops for an extended amount of time, resulting in a shortage of water. Therefore, people have learned how to regulate the amount of water in rivers. Dams are built across rivers and above the area where artificial lakes, which are called *reservoirs*, form downstream. Water from spate irrigation fills up dams and is then used throughout the year to irrigate fields, supply water to residential areas and generate power at hydroelectric power plants.

A spate, unlike a freshet, is when a river briefly and randomly overflows because of heavy rainfall or other weather disturbances.

So much precipitation falls in the Amur Basin in the rainy period of the year, for example, that rivers overflow.

Now we will answer two questions: Why don't rivers flow in a straight line? Why are strange horseshoe-shaped lakes often found near them?

Rivers flow on the Earth's surface because of gravity. However, even on plains they do not flow in a straight line; rather, they form bends called *meanders* (Fig. 6.5).

A Conundrum About a Cup of Tea Take a glass of boiling water and pour a spoonful of tea into it. Mix in the tea leaves completely (Fig. 6.6a) with a spoon and take it out. After the water has stopped spinning, the tea leaves will all gather at the bottom of the cup (Fig. 6.6b).

Why does this happen? Experience leads us to conclude that when mixing the tea and the leaves, the surface of the liquid is no longer flat and it bends. Let us assume that a liquid rotates in a vessel with the angular velocity ω. We will analyze a liquid cube with the mass Δm, which is located at the distance r from the axis of rotation (Fig. 6.7). Centripetal acceleration $a_c = \omega^2 r$ acts on the volume of liquid when it is rotating. This acceleration, as proven by Newton's second law, is equal to the difference of forces acting on the left and ride

Fig. 6.5 Meanders of the Büyük Menderes River in Turkey

Fig. 6.6 Cup of tea with tea leaves during and after mixing

sides of this cube:

$$\Delta m \omega^2 r = (p_1 - p_2) \Delta S. \qquad (6.4)$$

Here p_1 and p_2 is the pressure on the left and right sides of the liquid cube, and ΔS is the surface area of its side. Since $\Delta m \omega^2 r \neq 0$, then $p_1 \neq p_2$. Moreover, $p_1 = \rho g h_1$ and $p_2 = \rho g h_2$ where h_1 and h_2 is the distance from the left and right sides to the surface of the liquid mass.

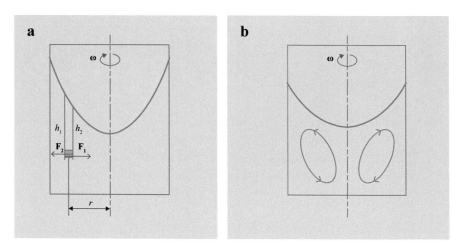

Fig. 6.7 Bending of the liquid surface and vortex flows inside the liquid

Since $p_1 > p_2$, it stands to reason that $h_1 > h_2$. Thus, the surface of the liquid mass will be curved (see Fig. 6.7).

Suppose we stop stirring the tea. The liquid will gradually stop spinning due to the friction force that is acting both between the liquid particles themselves and between the liquid particles and the walls of the cup.

Different conditions are acting on the layers of water on the surface and at the bottom of the cup. The friction between the water and the bottom of the cup is greater than the friction between the layers of the liquid itself and also between the surface of the liquid and the air. Therefore, the rotational speed of the water particles located at the same distance from the axis of rotation is different at different heights; in other words, it decreases with depth. This is why the particles near the bottom of the cup have such a small amount of centripetal force.

Consequently, a unique vortex flow results (Fig. 6.7b). The top layer of water moves faster and a stronger centripetal force acts on it than on the layer of water at the bottom.

Fluid flow carries the tea leaves toward the axis of rotation. When they connect by it, fluid flow sweeps them up and the heavier tea leaves clump up at the bottom of the cup.

The physical reasons that cause river channels to curve are the same as the ones that make the surface of water in a cup of tea curved. It is particularly remarkable that Albert Einstein was the first to present the physical analogy between curves in river channels and water movement in a cup of tea at a lecture that he gave at the Prussian Academy of Sciences in 1926.

Suppose that the river channel makes a turn. Centripetal acceleration caused by the turning of the outer bank will act on the water's surface. Therefore, this body of water will bend and rise up in the direction of the turn. In this case, an additional buoyant force is created. Figure 6.8 shows the forces acting on an element of liquid volume, the direction of the vortex flow of water and the velocity distribution for water along the depth of the channel. Water close to the river bottom experiences friction and slows down. Thus, the speed of the current is slower there than at the surface. Consequently, water at the surface rushes to the bank farthest from the river's turning point, that is, the outer bank, while water at the bottom, on the other hand, moves toward the turning point.

Because of the vortex flow, the bank that is far from the river's turning point collapses, while the one close to it, in contrast, receives more and more additional soil particles. This results in the river channel changing its course.

Water pressures causes the outer bank to recede, while the inner one washes out and rises up (Fig. 6.9). As a result, a small loop of the river initially begins to increase. We see tall cliffs on the outer bank, but gently-sloping beaches covered with sand and pebbles that are easy to access can be found on the inner bank.

If a river flows over rugged terrain, the shape of its channel is the same as the surface topography around it. Moreover, a river chooses a course that allows it to use a minimal amount of potential energy, which means that it flows along the deepest channel possible. However, if a river flows over a plain, it never runs in a straight line. Meanders form in any case when rivers

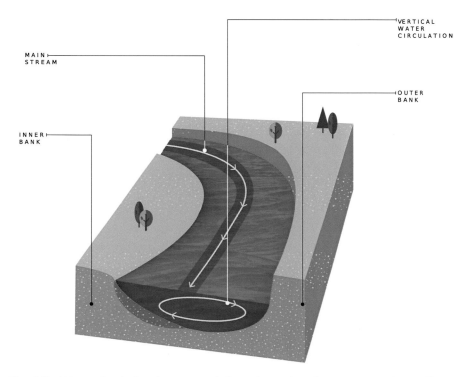

Fig. 6.8 Water circulation in a curved river channel and water's speed at different depths

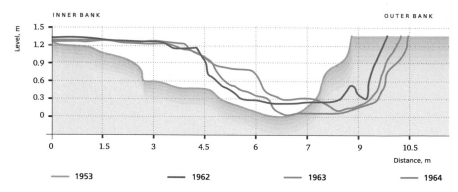

Fig. 6.9 Changes in the shape of a river's channel over time

randomly swing from side to side off their straight course with an increase in curves occurring as they move downstream.

In certain situations, a cutoff occurs at the base of a meander. The river channel then straightens out, and a loop known as an *oxbow lake* forms. After the river straightens out, it continues to flow along the new channel (Fig. 6.10), and this oxbow lake turns into a lake or fills up with mud and becomes a swamp.

Another specific tendency of the way river channels act on the Earth's surface is associated with the Coriolis force. As ones sees in Fig. 6.11, it is this force that is responsible for pulling water in rivers to the right bank in the Northern Hemisphere and to the left bank in the Southern Hemisphere.

Therefore, in the Northern Hemisphere, the right bank along which rivers flow is steeper and more eroded, while the left bank is flat. In the Southern Hemisphere, the opposite is true.

The physical reason for this phenomenon lies in the fact that the Coriolis force, which occurs due to the rotation of the Earth, pulls water to the right bank. However, friction causes the water flow velocity to be higher at the water's surface than at the bottom. This result is that a vortex flow, about which we have already learned, develops. It moves soil particles from the right to the left bank (Fig. 6.12).

We have analyzed the reasons why rivers are crooked. Now we will examine in greater detail the velocity of river currents and analyze the parameters on which it depends.

The velocity of a river's current is determined by the slope of its basin, as well as by the shape and roughness of its channel. It is evident that the greater the slope of the river basin in relation to a river's horizontal angle, the stronger the force of gravity that pulls a body of water downstream.

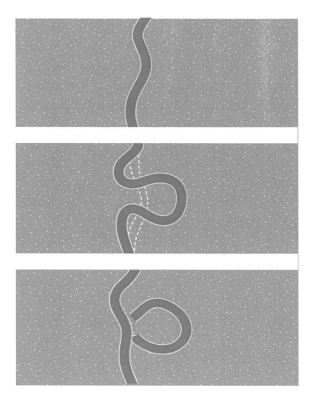

Fig. 6.10 Formation of an oxbow lake

Irregularities in the walls and riverbed of the channel cause water friction and slow it down. This means that the rougher a riverbed, the slower a river's velocity. For example, a riverbed made up of pebbles offers less resistance to water flow than one that is made of sand. Moreover, these irregularities cause the water flow to change from laminar to turbulent, which slows velocity down even more. The shape of the river channel also contributes to velocity. The smaller (for a given cross-sectional area) the area of the channel, the less friction there is, which means the river's velocity increases. It stands to reason that when a river channel has a semicircular shape, it covers a minimum area.

Now we will analyze how liquid flows in a river. The simplest case is the so-called *constant current* in which the velocity of a liquid at every given point remains constant both in magnitude and in direction. It is easy to imagine this type of current as many immiscible parallel streams flowing at the same velocity. Fluid particles come and go, but each particle that has just arrived at a specific point acquires the velocity that corresponds to that specific point. Therefore, by setting the fluid velocity at each point as a spatial coordinate,

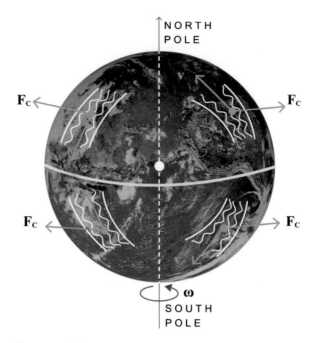

Fig. 6.11   The direction of the Coriolis force acting on water in rivers

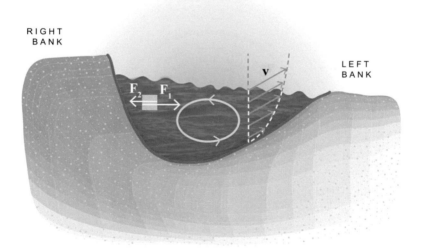

Fig. 6.12   The direction of water circulation, the forces acting on a fluid particle and the distribution of water velocity along the depth of a river's channel (Northern Hemisphere)

a constant current can be characterized as a velocity field, which is graphically depicted using flow lines (Fig. 6.13). In addition, the direction of vector **v** shows the motion direction of fluid particles at a given point, and the modulus illustrates the absolute value of velocity.

We will draw flow lines so that their density is proportional to the magnitude of velocity at a given point. The density or viscosity of the flow lines is the ratio of the number of lines ΔN to the size of the area ΔS through which they pass. This area is perpendicular to the flow lines.

We will call the part of the liquid that restricts the flow lines the *flow tube*.

Since we have a constant current, the location and shape of the flow tubes will not change over time.

Let's focus on a flow tube. Assuming that the cross-sectional area of the tube is very small, we can presume that the velocity of the liquid is the same at all the points in this cross section. The volume of liquid $V = Sv\Delta t$ passes through the arbitrary cross section S within the time Δt (Fig. 6.14a).

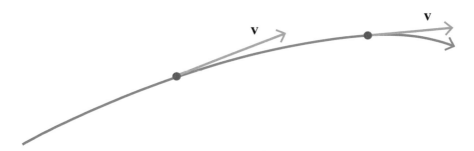

Fig. 6.13 Flow lines

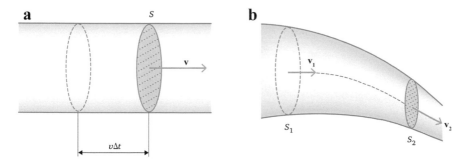

Fig. 6.14 A flow tube

We will choose two cross sections of the flow tube S_1 and S_2 (Fig. 6.14b). If the fluid is ideal and there are no fluid sources or outlets, then the volume of liquid that entered cross section S_1 will be equal to the volume that exited through cross section S_2. It follows that:

$$S_1 v_1 = S_2 v_2. \tag{6.5}$$

Since cross sections S_1 and S_2 have been chosen arbitrarily, it may be concluded that the variable Sv will be constant lengthwise along the entire liquid tube: $Sv = $ const. This is known as the *continuity equation*.

Hence, the flow rate increases at the spot where the tube narrows.

Anyone who has ever gone river rafting knows that as soon as a river narrows, the speed of its current increases. But on a wide stretch of river the opposite is true: sometimes the current practically disappears and rafters have to intensively row.

Bernoulli's principle explains the gravitational pull of bodies located near a flow boundary of moving liquids or gases. Sometimes this pull may be dangerous. When ships that are sailing on parallel courses come too close to one another, the speed at which water flows between their sides increases. Consequently, the amount of water pressure on the sides of ships in close proximity to one another becomes less than the external pressure, which can result in them colliding.

Bernoulli's Principle Let's focus on a flow tube restricted by cross sections S_1 and S_2 (Fig. 6.15) in a homogenous liquid that is flowing from left to right. We will assume that at the points corresponding to the cross sections S_1 and S_2, we know the velocities of the

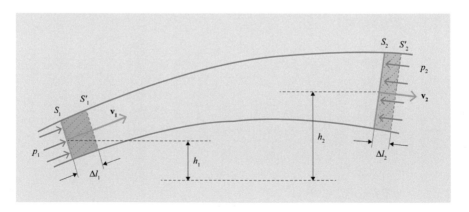

Fig. 6.15 Illustration of Bernoulli's principle

current v_1 and v_2, the fluid pressures P_1 and P_2 and the heights over the surface at which these points—h_1 and h_2—are located.

During the time Δt, the volume of liquid will move lengthwise along the flow tube. After moving along the trajectory Δl_1, let us assume that the cross section S_1 will move to position S_1'. Similarly, after having moved along the trajectory Δl_2, the cross section S_2 will move to position S_2'. Because of continuity of flow, the shaded areas will have the same value: $\Delta V_1 = \Delta V_2 = \Delta V$.

The energy of a moving fluid particle ΔV is equal to the total amount of its kinetic energy $\frac{1}{2}\rho v^2 \Delta V$ and to the potential energy in the gravitational field $\rho g h \Delta V$. Therefore, the energy flow rate that is running through the cross section S_1 within the time Δt will be equal to:

$$E_1 = \left(\frac{1}{2}\rho v_1^2 + \rho g h_1\right)\Delta V. \tag{6.6}$$

A similar formula for the energy flow through the cross section S_2 is:

$$E_2 = \left(\frac{1}{2}\rho v_2^2 + \rho g h_2\right)\Delta V. \tag{6.7}$$

A change in the total energy during flow movement $\Delta E = E_2 - E_1$ is equal to the external force energy that is performed by fluid friction against the walls of a vessel and by the pressure force on the side of the tube. Let's suppose that there is no friction; that is, we are dealing with a non-viscous liquid. Then only the work done by the pressure force will be different from zero. The work done by this force when applied to the cross sections S_1 and S_2 is equal to:

$$P_1 S_1 \Delta l_1 - P_2 S_2 \Delta l_2 = (P_1 - P_2)\Delta V. \tag{6.8}$$

The pressure force does not do any work on the side of the tube because of the fact that at every point it is perpendicular to the direction of the particles' movement.

We will make the change in the energy flow rate equal to external force energy. We get:

$$\frac{1}{2}\rho v_1^2 + \rho g h_1 + P_1 = \frac{1}{2}\rho v_2^2 + \rho g h_2 + P_2. \tag{6.9}$$

Since the cross sections S_1 and S_2 have been chosen arbitrarily, it is clear that the variable on the left and ride sides of the last equation will be constant for either cross section of the tube:

$$\frac{1}{2}\rho v^2 + \rho g h + P = const. \tag{6.10}$$

The formula we obtained is called *Bernoulli's principle*. This equation expresses the principle of conservation of mechanical energy when there is a constant current of incompressible non-viscous liquids. Since a current flows faster where a river narrows, according to Bernoulli's principle, pressure decreases there.

The same thing happens when a vehicle is traveling at a high speed. All objects around it begin to be pulled toward it. This is easy to see in fall when leaves fall on the ground because they seem to be pulled in the direction of passing cars and "run" after them for a little bit as if they want to catch up to them. Turbulent air flow, which swirls around moving cars and push leaves forward, is at work here. If a very long vehicle is moving extremely quickly, this situation may even pose danger for people. When the Russian high-speed train *Sapsan*, for example, flies at a speed of more than 200 km (124.27 mi)/h, rarefaction waves create a real danger for people on the train platform who could fall underneath the train.

Therefore, there is a safety line on train and subway platforms for a reason: it is deadly to stand in front of it at the edge of a platform when a train is moving!

6.3   Waterfalls

Everyone knows that if a river falls from the edge of a cliff, it is a beautiful sight. This is called a *waterfall* (Fig. 6.16a). But in order to be able to rightly call itself a *waterfall*, a body of water must meet two criteria. First, the water must fall at an angle of more than 45° (otherwise this type of natural occurrence is known by a much less romantic name—*falls* [Fig. 6.16b]). Second, the height from which the water falls must be over 1 m (39.37 in) (otherwise this is called *rapids* [Fig. 6.16c]). If the water falls over several ledges, then a series of waterfalls forms, which is known as a *cascade*.

The main characteristics of waterfalls are: height (i.e., the height from which water plummets down), width (i.e., the width of the water at the crest) and stream flow rate (i.e., the rate at which the water flows over the crest).

There are various reasons why waterfalls form on rivers. A waterfall ledge, for example, can form because rocks break, whereby one large slab becomes both horizontally and vertically displaced in relation to another. In addition, steep steps can form in the river channel as a result of the river causing rocks of different strengths to erode. Lastly, the river channel may be blocked by a landslide, an avalanche or a glacier, which, in turn, causes a lake to form, the outlet of which flows over a waterfall ledge.

The cascading streams of water do a tremendous amount of work, which is evident in the large volume of falling water that dislodges rocks at the base of a waterfall. After this happens, the more massive rocks upstream erode. In this way, little by little waterfalls retreat upstream. This is shown happening

Fig. 6.16 The difference in vertical drop on a river: **a** a waterfall on the Shinok River (Altai Republic), **b** Uchar Waterfall[1] (Altai Republic), **c** rapids on the Katun River (Altai Republic)

in Fig. 6.17 where Niagara Falls serves as an example. The average rate that the Falls migrate upstream is approximately 1 m (39.37 in)/year.

When looking at a large waterfall, on what should we focus our attention besides the waterfall itself?

When we see pictures of the highest waterfalls, it is possible to spot a cloud of mist that forms at a waterfall's base. The reason this happens is because streams of water split into drops, which then split into tiny droplets. This causes a mist to form (Fig. 6.18).

Let's consider how drops of water develop and how high they can fly up after hitting the water's surface.

---

[1] In English this is referred to as *Uchar Waterfall*, while in Russian it is known as *Uchar Falls* (translator's note).

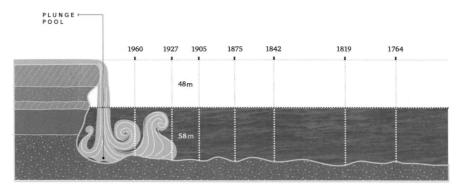

Fig. 6.17 Changes in Niagara Falls' (Canadian side) movement over the past 300 years

Fig. 6.18 Angel Falls, Venezuela

First, it is important to note that when a spray of water falls from a high altitude, the process very quickly begins by which a continuous flow of water breaks up into individual drops. In fact, as soon as the drops free fall off a waterfall ledge, they begin to move with increased speed. In addition, those sections of water that fall down earlier move at a higher speed that those that follow at each particular moment afterward. As a result, the spray of water is stretched until it breaks up into individual drops.

When water falls into the plunge pool at the base of the waterfall, drops continue to break up into tiny droplets. Moreover, a large number of small droplets or splashes (Fig. 6.19) are formed that subsequently fly up because of ascending air masses.

Thus, a drop of water that falls from a height of 200 mm (7.87 in) creates a splash that is 25 mm (0.98 in) high; in other words, about 10 times less than its height. However, if we use the height of the waterfall in place of 200 mm (7.87 in), we will not get this kind of correlation. This is due to the fact that a drop that falls from a high altitude moves at an accelerated speed only at first; thereafter, when the force of gravity becomes equal to air resistance, its speed does not fluctuate.

Height of Water Splash Let's assume a drop of water with the radius r falls on the water's surface at the height h.

If we disregard air friction, then the drop's kinetic energy at the moment it makes contact with the water's surface is equal to:

$$E_k = \frac{4\pi}{3}\rho_{water}gr^3h. \tag{6.11}$$

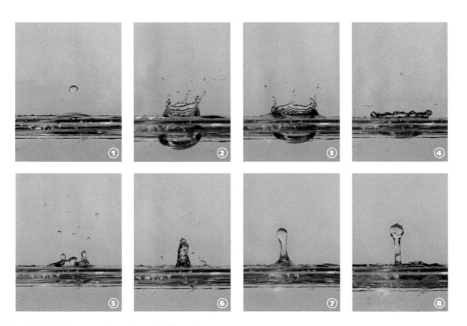

Fig. 6.19 Phases of a drop falling into water

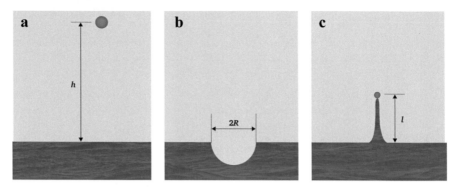

Fig. 6.20 Phases of a drop of liquid falling on a liquid surface

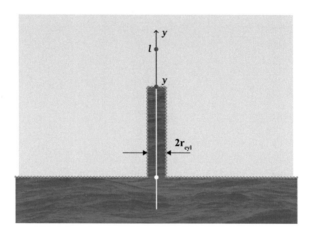

Fig. 6.21 The model of the liquid splash

Before the drop hits the water's surface, it possesses not only kinetic energy, but also energy connected with surface tension: $E_{surf} = 4\pi r^2 \sigma_{water}$. This energy is significantly less than kinetic energy, and it can be ignored. Kinetic energy E_k is first spent on a making depression with the radius R that appears in a liquid and then on a splash with the height l (Fig. 6.20), which we will simulate in the shape of a cylinder with the base $2r_{cyl}$ (Fig. 6.21).

In order for a splash of this shape to form, effort must be exerted against the force of gravity and against surface tension because the liquid's surface area increases. The effort exerted against the force of gravity is equal to:

$$A_g = \frac{1}{2}\pi\rho_{water}gr_{cyl}^2l^2. \tag{6.12}$$

The effort exerted against surface tension is proportional to the increase in the surface area of the liquid $2\pi r_{cyl}l$ and is equal to $A_{surf} = 2\pi\sigma_{water}r_{cyl}l$. According to the law of conservation of energy, $E_k = A_g + A_{surf}$ or $\frac{1}{2}\pi\rho_{water}gr_{cyl}^2l^2 + 2\pi\sigma_{water}r_{cyl}l - \frac{4\pi}{3}\rho_{water}gr^3h = 0$.

This is a quadratic equation with respect to the variable $r_{cyl}l$. Its solution is written as:

$$r_{cyl}l = \sqrt{\frac{4\sigma_{water}^2}{\rho_{water}^2 g^2} + \frac{8}{3}r^3 h} - \frac{2\sigma_{water}}{\rho_{water}g}. \tag{6.13}$$

After finding that $h = 200$ mm (7.87 in) and $r = 2$ mm (0.078 in), we get: $r_{cyl}l = 52$ mm^2. The radius of the base of the splash is usually approximately equal to the radius of the falling drop. Therefore, we can confirm that $r_{cyl} \approx 2$ mm (0.078 in) and $l \approx 25$ mm (0.984 in).

Yet another reason why drops break up is because they become deformed by a countercurrent. Some large drops that fall from a high altitude become deformed and break up because of it (Fig. 6.22).

The criterion according to which a drop of water can exist as a whole is determined by the fact that its energy, which is related to surface tension, exceeds the amount of energy that is contingent upon air resistance when the drop if falling. If a drop of water with the diameter d falls at the speed v, the viscous force from air resistance equals $\frac{1}{2}\rho_{air}v^2$. The surface tension is $\frac{\sigma_{water}}{d}$ (σ_{water} is the coefficient of the surface tension of water; at room temperature, it is equal to 7.2×10^{-2} N/m).

Thus, drops will break up provided that the speed at which they fall exceeds a certain critical value, which is:

$$v \geq v_{crit} = \sqrt{\frac{\sigma_{water}}{\rho_{air}d}}. \tag{6.14}$$

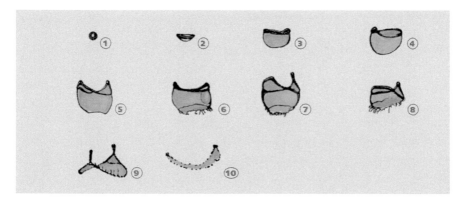

Fig. 6.22 Deformation and break up of drops by a countercurrent

By substituting the values of the physical parameters of water and air and finding that $d = 5$ mm (0.196 in), we get: $v_{crit} = 3.5$ m (11.482 ft)/s.

Now let's estimate what the speed of a drop is when it falls from the height of a waterfall, which is, for example, 100 m (328.084 ft). If we disregarded air resistance, then the speed would be equal to $v = \sqrt{2gh} = 44$ m (144.357 ft)/s. However, in reality air resistance cannot be disregarded, at least not for large drops. Thus, the speed of the drop will be considerably less. When calculating the speed of drops with a diameter of about 5 mm (0.196 in), it is necessary to take into consideration the turbulent air flow around them. We will discuss this in more detail when we analyze the shape of raindrops. In this situation, the speed of a drop is determined by the formula:

$$v \approx \sqrt{\frac{g\rho_{water}d}{\rho_{air}}} = 6.4 \text{ m } (20.997 \text{ ft})/\text{s}. \qquad (6.15)$$

This is approximately the speed that is experimentally observed when large drops of water fall from a height of about a hundred meters (feet). This is higher than the critical speed, which causes the drops to break up and produce numerous splashes of water.

When liquid sprays, drops become electrostatically charged. This is called the *Lenard effect*.

A water molecule is a dipole. The dipole moment of a water molecule **p** is located on the bisector of the H–O–H angle (Fig. 6.23).

When a large drop forms, the negatively charged "tails" of the outer water molecules protrude from it, which creates a negative charge on the drop's surface and the positively charged "tails" hide inside of it (Fig. 6.24) As a result, a thin film forms on the drop's surface, the outer side of which has a negative charge, while the inner side has a positive charge. This gives way to the development of an electric field that affects free ions both inside and outside of the drop. Experiments have proven an interesting fact: if these drops break up, the charges of the smaller drops that form will not be zero.

Small splashes, which are formed at the base of the waterfall during jet breakup, have a predominantly negative electric charge, while large splashes have a positive charge. Ascending air masses can move small splashes of water to altitudes that are as high as several dozen meters (hundreds of feet). Large drops settle to the bottom of these air masses. Thus, an electric field develops that is directed upward, the intensity of which can reach a very significant

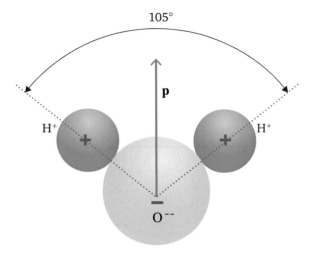

Fig. 6.23 Structure of a water molecule

Fig. 6.24 Diagram showing how water molecules are positioned at the surface of a large drop of water

magnitude—several tens of volts per centimeter (inch). The more powerful the waterfall, the stronger this effect is. For example, an electric field with a strength of about 25 kV/m is found not far from Victoria Falls on the Zambezi River in South Africa. It is 133 m (436.351 ft) high, 1600 m (5249.343 ft) wide and exists because of the Lenard effect. This is a very high electric charge (the electric field of the Earth's surface is approximately 0.15 kV/m). People are not advised to spend more than 10 min in areas such as these where there are such high electric charges.

We have already stated that there is a tremendous amount of energy in the water that flows in a waterfall and since ancient times people have been adept at using hydraulic power. A device that converts hydraulic energy into mechanical energy is called a *waterwheel* (Fig. 6.25a). It was connected to millstones in a mill so that they would turn and grind grain. The entire integrated system is called a *watermill* (Fig. 6.25b). It originated much earlier than the windmill.

Despite the fact that today's modern-day technology used by hydroelectric power plants to capture energy from falling water (Fig. 6.26) bears little resemblance to watermills, the principle behind their operation is the same. The fact is simply that waterwheels have turned into today's hydraulic turbines and the mechanical power that was once generated has changed to hydroelectricity.

Capacity of a Run-of-the-River Hydroelectric Power Plant When we were deriving Bernoulli's principle, we obtained an equation for the energy flow of a liquid. This equation makes it possible to estimate the energy potential of a specific river. The stream flow rate is equal to:

$$W = \left(\frac{1}{2}\rho v^2 + \rho g h\right)\Delta V, \tag{6.16}$$

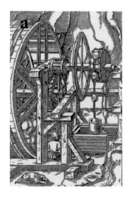

Fig. 6.25 Using water energy: **a** a waterwheel powered by a pump (an illustration from Georgius Agricola's book *De Re Metallica Libri XII*, 1556), **b** a working watermill at a museum in Bugrovo Village (Pskov Region, Russia)

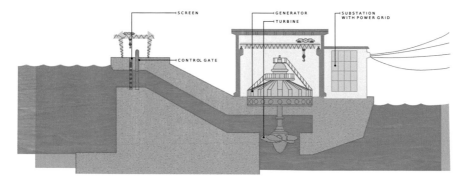

Fig. 6.26 A diagram of a modern-day hydroelectric power plant

Fig. 6.27 Sayano-Shushenskaya Hydroelectric Plant on the Yenisey River, Russia

where ΔV is the volume of water that flows through a cross section of the stream in a unit of time. In hydraulics, this value is called *flow rate* Q and is measured as m^3/s. The most efficient hydroelectric power plants are those that use water's potential energy (the second item in the equation). In order to increase elevation, dams are built and reservoirs and weirs are constructed. If we disregard the input of kinetic energy, we can estimate the capacity of a hydroelectric power plant this way:

$$W = \eta \rho g h Q. \tag{6.17}$$

Here η is the efficiency coefficient of the turbine, which for modern plants is approximately 0.80–0.90 and h is the elevation between the level of the reservoir and the level at which the turbine is located. If we insert, for example, $h = 10$ m (32.808 ft) and $Q = 10$ m^3/s, we get $W \approx 1$ MW from one turbine. Today's hydroelectric power plants have a much higher capacity. For example, the Sayano-Shushenskaya Hydroelectric Plant on the Yenisei River (Fig. 6.27) has a capacity of 6400 MW! Water that flows from a height of 194 m (636.482 ft) turns ten turbines, each of which uses 358 m³/s. When looking at such a powerful structure, it is hard to believe that it was made by human beings.

6.4   Glaciers and Icebergs

In high latitudes (or high in the mountains), a significant amount of snow falls in winter and summers are short and cold. The snow that has fallen in winter does not always have time to melt in summer. It then begins to accumulate and grow in size. Under the weight of new masses of snow, the lower layers get crushed, packed down and turn into ice. In this way, after hundreds, thousands and millions of years amazing creations of nature are formed, which are known as *glaciers*. Glaciers are so fascinating and complicated that an entire branch of science called *glaciology* was developed to study them.

Today glaciers occupy about 11% of land surface. This is an extremely large area—about 16.1 million km². Therefore, it is not surprising that a tremendous amount of freshwater is found in glaciers—more than 26 million km³ (6.23773 mi³). This is almost 69% of all of the Earth's water supply. The amount of water contained in glaciers corresponds, for example, to the overall amount of atmospheric precipitation that falls on the Earth in 50 years and also to the amount of water that flows in all rivers in 100 years!

The largest glaciers are located in Greenland and Antarctica. The total volume of ice in Greenland is 2.6×10^6 km³, while there is even more ice in Antarctica— 24.2×10^6 km³. The size of the Antarctic Ice Sheet is so massive that it pressed the continent into the Earth's mantle with its weight so that in some places the level of the ice–rock interface is now lower than the level of the ocean. If an increase in temperature causes the ice in Greenland to melt, the level of the World Ocean will rise 7 m (22.965 ft). But if the ice in Antarctica melts, the level of the World Ocean will rise 70 m (229.659 ft) and then we will all be in quite a bit of trouble!

What makes glaciers so intriguing? A mountain glacier is, for all intents and purposes, an ice river (Fig. 6.28) cascading down a mountain. The speed at which a glacier moves depends on its mass, bottom slant, which is the glacier bed, temperature and whether there is water in it. Qualitatively, we understand that the higher the bottom slant, the temperature of ice and the

Fig. 6.28 Serrano Glacier (Torres del Paine National Park, Patagonia, Chile)

larger the glacier's mass, the faster the speed at which it travels downhill. This speed can vary over a wide range from 0.1 m (0.328 ft) to 50 m (164.042 ft)/day.

The phenomenon of ice formation in glaciers is quite fascinating.

How does snow end up on a glacier bed? It either falls as frozen precipitation or it is brought from surrounding slopes by avalanches. Snow can remain on level and flat areas for as long as 100 years. The sun, wind and thaw cycles cause snowflakes to change their shape: they lose their star-like pattern and turn into granular snow. When the sun warms up a snowy slope, snow melts and water seeps into it. Water then freezes, which makes the crystals there increase in size. The sublimation of water vapor, which freezes onto the crystals, also plays a part in their growth.

> Snow that has a granular structure and is more than one year old is called *firn*.

The bits of granular snow that make up firn grow slowly and reach several millimeters (inches) in size. It should not come as a surprise that the older firn is, the larger the size of its granules.

When these granules grow, air is pushed out of inter-granular spaces. When the air passages between the granules completely disappear, they stick together and firn ice forms, which is a dense white material (Table 6.1). For the most part, this is the type of ice that forms on city streets if snow is not shoveled. Pedestrians and cars pack snow down and after several days it turns into solid

Table 6.1 Densities of different types of snow, firn and ice that make up glaciers

Snow/ice	Density, kg/m³
New dry snow	50–70
Damp newly-fallen snow	100–200
Settled snow	200–300
Wind-packed snow	350–400
Firn	400–830
Glacier ice	830–916

ice. It is very difficult to chip this type of ice off sidewalks. Firn ice forms much slower in nature and it happens over a long stretch of time. Due to the fact that layers of firn exert pressure on lower layers of white firn ice, it turns into glacier ice. A fifty-meter layer of firn (164.042 ft) turns into ice after 50 years. But if a layer of firn is less than 50 m (164.042 ft), the layers above it do not exert enough pressure on it for dense ice to form. This type of "failed" glacier is called a *snowfield*.

Ice is an unusual substance. It is both fragile and pliable. Moreover, the higher the temperature and pressure, the more flexible it is. If constant pressure is put on ice, it will start to yield under this strain and slowly begin to creep. In this same way, layers of firn exert pressure on glacier ice, thus forcing it to the surface. In which direction will this ice move? Obviously in the direction where it will be carried away by the force of gravity, which is down a slope. The steeper the inclination of the slope, the less pressure is required for glacier ice to begin to flow downslope. A 2-m-thick (6.56 ft) layer of firn is enough to make glacier ice slide down a slope at a 40–45° angle. However, a 60-m-thick (196.85 ft) layer of firn is needed to make glacier ice flow down a gentle slope with an inclination of several degrees. Studies of glaciers have shown that in many respects glacier movement is very similar to laminar flow. Ice friction against the glacier's outer edges and the glacier bed makes ice in the middle of a channel flow faster than by the edges of a river. In addition, ice on the surface of water travels faster than when it is on the riverbed. When the glacier bed curves, the flow line, which corresponds to maximum flow velocity, shifts to the bank farthest from the river's turning point. Finally, if the glacier bed narrows, glacier flow increases.

If you put an even row of small stakes on the surface of a glacier, over time they will bend as shown in Fig. 6.29 where ice moves at different velocities along the glacier's cross section.

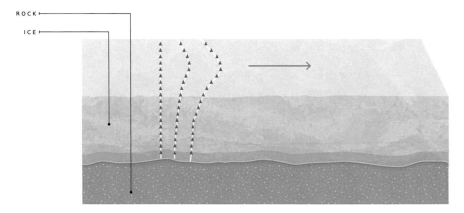

Fig. 6.29 An experiment to measure the rate of ice creep

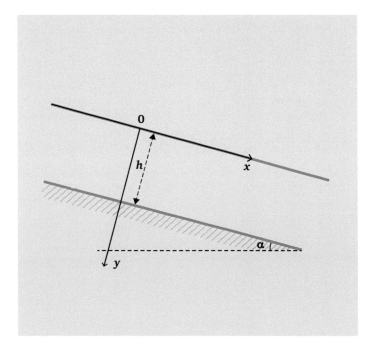

Fig. 6.30 The simplest model of glacier flow

Glacier's Thickness and Rate of Flow Let's consider the simplest model of glacier flow. We will assume that an ice sheet with parallel sides and the thickness h lies on a rough plane with the tilting angle α. We will also assume that the plate does not slide along the base plane; that is, its downward motion is only due to deformation that comes from its own force of gravity.

We will analyze a column of ice with a single cross section (Fig. 6.30). The pressure on its base that is created by gravity along the cleavage plane is $\rho_{ice} g h \sin \alpha$. This pressure is balanced by the friction force p, which is applied to the base of the ice column. Hence:

$$p = \rho_{ice}\, gh \sin \alpha. \qquad (6.18)$$

The variable p was calculated using a formula for different glaciers when they have various values for h and α and it proved to be in the range of 50–150 kPa. Glacier ice is generally viewed as a pliable substance with the yield strength $p_0 = 100$ kPa. Yield strength is as important for pliable substances as tensile strength, which is the maximum force a material can withstand and still retain its shape, is for fragile ones.

Thus, in this model glacier motion occurs because the pressure on its lower edge reaches the yield point. Therefore, it is possible to write the formula as:

$$h = \frac{p_0}{\rho_{ice}\, g \sin \alpha}. \qquad (6.19)$$

Based solely on the results we obtained from measuring the surface slope, this formula allows us to calculate a glacier's thickness. We see that in places where there is a slight surface slope, glaciers will be thicker, while in areas with a steep surface slope, they will be relatively thin. This conclusion has been confirmed by experimental observations.

The second circumstance that causes glacier motion is sliding on the glacier bed when ice reaches the melting point. In this case, we will assume that the resistance to glacier motion is caused by protrusions in the glacier bed. On the sides of the protrusions, which face upwards toward the slope, there is excessive pressure, which causes the ice to melt. The water that has built up flows around a rise in the ground and refreezes on the opposite side where the pressure is lower. We will assume that the ice and the glacier bed are separated by a film of water. The glacier bed itself is an inclined plane with protrusions shaped like cubes with sides denoted as a and a distance between the cubes denoted as l. If p is the pressure force on a unit area at the cube's base that is created by the friction force as it was earlier, then since one cube has the area l, the friction force attributable to each cube will be equal to pl^2. This force is channeled up the slope of the cube face (with an area of a^2) and creates pressure that is $\frac{pl^2}{a^2}$. Pressure on the top of the cube causes a change in the melting point of ice relative to the cube's bottom by the following amount:

$$\Delta T_m = Cp\left(\frac{l}{a}\right)^2, \qquad (6.20)$$

where $C = -7.4 \times 10^{-5}$ K/kPa.

Let us assume that u is the speed of glacier flow. Then a volume of ice equal to ua^2 melts in a unit of time on the edge of rise in the ground that is facing upstream. The water that has built up flows around it and refreezes once again by its lower edge. When freezing such as this occurs, a certain amount of heat $ua^2 \rho_{ice} \lambda_{ice}$ ($\lambda_{ice} = 3.35 \times 10^5$ J/kg is the heat of fusion of ice) is released, which is transferred to the top of the rise and used to melt ice. The amount of heat that is transferred in a unit of time will be equal to $\kappa_b a \Delta T_m$, where κ_{tc} is the thermal conductivity coefficient of bedrock, which makes up the glacier bed.

We get:

$$ua^2 \, \rho_{\mathrm{ice}} \, \lambda_{\mathrm{ice}} = \kappa_{\mathrm{b}} \, a \Delta T_{\mathrm{m}}. \tag{6.21}$$

By substituting ΔT_{m} from the formula shown above, we find the glacier's rate of flow:

$$u = \frac{C \kappa_{\mathrm{b}} p}{\rho_{\mathrm{ice}} \, \lambda_{\mathrm{ice}} a} \left(\frac{l}{a} \right)^2 \tag{6.22}$$

The parameter l/a characterizes the roughness of the glacier bed's surface. Based on this formula it follows that the smoother the glacier's surface, the faster it travels.

Let's take off on a trip with ice (Fig. 6.31). Snow fell and turned into densely packed snow, that is, firn. While the snow is becoming compact, it moves and turns into ice, but the ice is heavier and moves faster so it breaks off the firn. A tremendous crevasse forms where the moving glacier separates from the firn. This crevasse is so deep (up to 150 m [492.126 ft] deep and 30 m [98.4252 ft] wide) that in summer its rock bed is visible. This kind of crevasse is called a *bergschrund*.

The glacier then slides down the glacier bed and sometimes runs into a steep mountain slope. If this had been an actual river, a waterfall would have formed. But the ice forms vertical pinnacles or columns of glacier ice known as *seracs*. Seracs are often unstable and because they are so unpredictable, they can be very dangerous for mountain climbers.

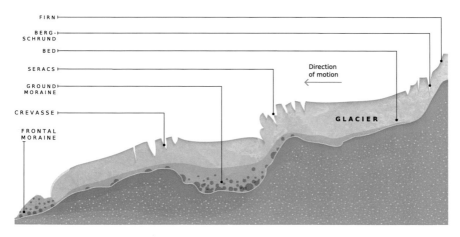

Fig. 6.31 Cross section of a glacier

Lastly, the area where a glacier stops is called the *frontal (or end) moraine*. Rock debris—from small cobblestones and clay to giant slabs up to several hundreds of meters (tens of feet) in diameter—mixed with ice is found in this very unwelcoming place. The area that consists of the debris that is transported by the glacier's bottom layer is called the *ground moraine*.

Thanks to glaciers icebergs form (Fig. 6.32).

> An iceberg is a floating block of ice that broke off a glacier and towers more than 5 m (16.4042 ft) above sea level.

These slabs of ice may be larger in area than some countries (Fig. 6.33)!

The largest number of icebergs come from the Antarctic Ice Sheet. About 100,000 icebergs constantly float close to Antarctica. Most icebergs in the Northern Hemisphere come from Greenland. Each year up to 15,000 ice blocks break off it.

What causes huge chunks of ice to break off the Antarctic Ice Sheet? A small amount of thawing occurs at the edge of the ice shelf, which results in the ice becoming thinner. However, when a glacier is a significant distance from the water's edge, the reverse occurs: ice accretion is at work on the bottom of it.

Fig. 6.32 An iceberg by the shores of the Antarctic

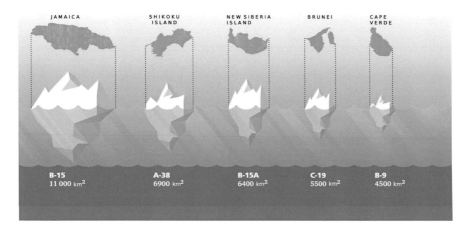

Fig. 6.33 Comparison of the largest icebergs in relation to islands

Ice accretion develops as a result of a decrease in the salinity of water underneath the glacier because melting snow dilutes ocean water with fresh water. Fresh water on the bottom of a glacier freezes over at a more intensive rate, which results in the formation of an unusual cross section of an ice slab. In the area where the freezing occurred, the buoyant force (as per *Archimedes' principle*), which pushed up this slab of ice, is significantly higher than in the spot where the glacier broke off (Fig. 6.34). The pressure that has built up in the ice causes it to break and separate.

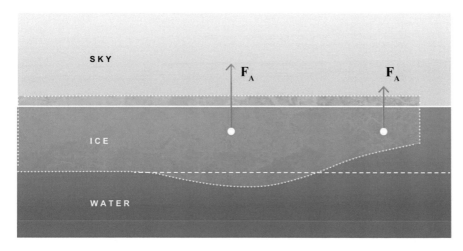

Fig. 6.34 Buoyant force acting upon different areas of the ice shelf

Fig. 6.35 The largest portion of an iceberg is submerged underwater

The slab of ice that broke off becomes its own separate entity—an iceberg (Fig. 6.35). Part of it towers above the water and part of it lies deep below it.

Ratio Between the Tip and the Submerged Section of an Iceberg Let's assume an iceberg has the volume V. We will denote the volume of its tip as V_{tip} and the volume of its submerged section as V_{sub}. It goes without saying that $V = V_{tip} + V_{sub}$. We will accept that the ice density is equal to $\rho_{ice} = 920 \ kg/m^3$, while the sea water density is equal to $\rho_{s.w.} = 1025 \ kg/m^3$. The iceberg's equilibrium condition is determined by the equilibrium of the force of gravity acting upon the entire iceberg and the buoyant force, which acts upon its submerged section.

$$\rho_{ice} \left(V_{tip} + V_{sub} \right) = \rho_{s.w.} V_{sub}. \tag{6.23}$$

From here it follows that:

$$\frac{V_{sub}}{V_{tip} + V_{sub}} = \frac{\rho_{ice}}{\rho_{s.w.}} = 0.90 \tag{6.24}$$

Hence, 90% of an iceberg's volume is submerged below the ocean's surface. If an ice slab floats in fresh water, the volume of its tip will be even less—about 8%.

Icebergs caught in ocean currents begin their journey around the World Ocean and sometimes even end up at the Equator. They gradually melt, of course, and as they do so, they release millions of tons of absolutely fresh water into the ocean. In addition, because of the difference between the temperature of icebergs and the environment they can be used to power heat engines.

How Much Energy Is Available from a Melting Iceberg? A tremendous amount of energy is needed for ice to melt. The heat of fusion of ice is $\lambda_{ice} = 3.35 \times 10^3$ J/kg. Let's assume an iceberg with the mass $M = 10^{10}$ kg is drifting along the Gulf Stream. We will estimate how much work a heat engine can perform before the iceberg completely melts if this engine uses water from the Gulf Stream as a heater and an iceberg as a refrigerator. We will accept that the temperature of the iceberg is $T_i = 0\,°C\,(32\,°F) = 273$ K and the temperature of the water in the Gulf Stream is $T_G = 22\,°C\,(71.6\,°F) = 295$ K. Then the required amount of work will be equal to:

$$\lambda M \times \frac{T_G - T_i}{T_G} = 2.7 \times 10^{14}\,\text{J}. \qquad (6.25)$$

This is approximately the amount of energy that a hydroelectric power plant with a capacity of 1 million kW can produce within the span of three days' time.

In the waters of the Southern Hemisphere, a cracking sound may be heard underwater. These are the sounds of Antarctica and simply the cracking noise that icebergs make when they are melting.

Every time an iceberg cracks it generates a sound wave with a frequency of about 10 Hz, which propagates without subsiding for thousands of kilometers (hundreds of miles). Since there are many icebergs and they are continuously melting, a constant underwater cracking sound, which becomes especially loud in the summer months, is heard in the southern part of the World Ocean.

Volume of an Iceberg's "Song" First, we will estimate the amount of energy released when a single crack appears in an iceberg. Then we will estimate how many such cracks can appear in a large iceberg. Finally, we will determine the intensity of sound waves in the ocean.

This cracking sound is a sound wave produced at the moment that a crack appears, i.e., when an iceberg breaks. The acoustic energy that is released when this break occurs is proportional to the surface area of the break S and to the variable σ, which is a material property. The variable σ is measured as J/m^2 and has the physical meaning of acoustic energy, which is released when a break with an area of 1 m^2 appears. We will analyze a sample rectangular piece of ice with the length L and the cross-sectional area S. If its length increases to ΔL when pressure is applied, the energy increment will be equal to:

$$\Delta Q = \frac{1}{2}ESL\left(\frac{\Delta L}{L}\right)^2 = \sigma S. \qquad (6.26)$$

Here E is Young's modulus of polycrystalline ice, which is approximately equal to 7 GPa. The tensile strength of ice σ_{TS} is about 1 MPa. If we define the length of this ice sample as $L = 1$ m (3.28 ft), the percentage elongation of it is $\frac{\Delta L}{L} \approx 10^{-4}$. After making some substitutions, we get $\sigma \approx 30$ J/m^2. But it is important to take into account the fact that this ice, which makes up an iceberg, came from a glacier and has a tremendous number of imperfections. This means that its stress–strain properties are far inferior to the corresponding values for polycrystalline ice. Bearing this in mind, we will accept that $\sigma \approx 1$ J/m^2.

Now we will estimate the overall fracture planes that emerge when a large iceberg breaks up into smaller pieces.

We will start with a very large iceberg that has the thickness d and the linear dimensions $R \gg d$, and we will break it in half by cracking it (Fig. 6.36). The first break down the middle creates a crack with the surface $S \approx Rd$. The second break affects both sides of each half and makes a crack that has the same total surface area as the first one.

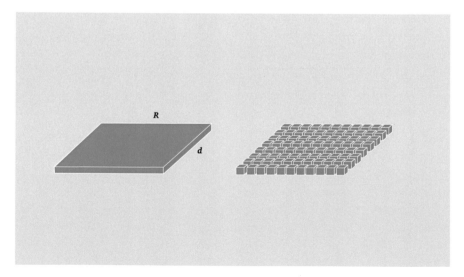

Fig. 6.36 In calculation of the total area of breaks because of iceberg cracking

The third break covers twice as large of an area as the first two did. And so the process continues this way. The greater the number of small icebergs that are formed, the larger the total area of the new crack will be when the next break splits the pieces in half. For the purposes of our experiment, it makes sense to break the iceberg into pieces the same way we already did until the size of each one matches the thickness of the iceberg. If an iceberg's linear size is equal to its thickness, then since a crack has formed, the break in it will most likely remain above the water's surface. This means that this break will not have anything to do with the acoustic effects made underwater. Using this type of approximation, the total length of the breaks will be $2R \times \frac{R}{d}$ and the total area will be $S = 2R^2$.

It is well known that Antarctica sheds about 10^{15} kg of ice each year in the form of large icebergs. If we shatter them all into pieces of about $d \approx 200-300$ m (656.168–984.252 ft), we create breaks that have a total area of approximately $3 \div 5 \times 10^9$ m^2. In this case, the amount of general acoustic energy released will be equal to $Q_{total} \approx 4 \div 5 \times 10^9$ J.

Now we will estimate the intensity of sound waves that have developed in the ocean. Acoustic energy, which corresponds to the cracking of icebergs, is continuously emitted into the ocean during the warm season and is equal to approximately 10^7 s. Some of this energy is lost (i.e., it is absorbed by the bottom of the ocean, is re-reflected or radiates toward Antarctica), while some of it travels to the open ocean in the north and creates those very same "sounds of Antarctica." The intensity of sound waves (i.e., the amplitude per unit area) in moderately high southern latitudes can be estimated in the following way. We divide the energy radiated in this direction by the entire time frame during which the sound is emitted and by the area of the imaginary line that covers all of Antarctica along, for example, the 60th parallel and several kilometers (miles) wide. This area will be about 10^5 km. By substituting numerical values, we will get the sound energy flux:

$$I \approx 10^{-9} \text{ W/m}^2. \tag{6.27}$$

The last step is to convert everything to decibels. Sound pressure, which is included in this formula for volume, is the amplitude characteristic that shows to what degree pressure deviates from the average. Intensity is the energy characteristic of a sound wave and is expressed in terms of the square of the amplitude. The formula used to calculate intensity and sound pressure looks like this:

$$P^2 = I\rho_{water}c, \tag{6.28}$$

where $c = 1500$ m/s (4921.26 ft) is the speed of sound in water.

Hence, the intensity that was found corresponds to about 0.04 Pa of sound pressure.

The decibel (dB) unit for measuring volume is a logarithmic scale that shows how much the pressure in a sound wave (P) exceeds base pressure P_0. The measurement of volume in decibels is 201 g (P/P_0). It is customary to use $P_0 = 1\,\mu\text{Pa}$ as the base pressure in underwater acoustics. This results in the final estimate of 90 dB for volume. To put this into perspective, this is the loudness of a freight train 7 m (22.96 ft) off in the distance.

Further Reading

1 Bentley, W.A.: Snowflakes in Photographs. Dover Publications (2000)
2 Cuffey, K., Paterson, W.S.B.: The Physics of Glaciers, 4th edn. Academic Press (2010)
3 Dingman, S.L.: Physical Hydrology, 3rd edn. Waveland Press (2015)
4 Fleming, S.W.: Where the River Flows: Scientific Reflections on Earth's Waterways. Princeton University Press (2017)
5 Green, H.L., Lane, W.R.: Particulate Clouds, Dusts, Smokes & Mists. E. & F. N. Spon Ltd., London (1957)
6 Julien, P.Y.: River Mechanics, 2nd edn. Cambridge University Press (2018)
7 Van Dyke, M.: An Album of Fluid Motion, 14th edn. Parabolic Press, Inc. (1982)
8 Varlamov, A.A., Aslamazov, L.G.: The Wonders of Physics, 4th edn. World Scientific (2019)

7

The Weather and Climate

Abstract In this chapter, we discuss the physical background beyond the concepts of climate and weather. First, we study the factors affecting the climate at a given point on the Earth's surface. Then we describe such weather phenomena as clouds, fog, rain, snow, and hail. We also review the manifestations of atmospheric electricity, such as thunder and lightning. The mechanisms of snowflakes formation and their types are discussed, as well as such phenomena as blizzards and snow avalanches in mountains. In conclusion, we examine the astronomic factors affecting global climate change on the Earth and explore the physical processes that explain why the Earth's surface warms up and cools down over time.

Every day when we look out the window in the morning and go outside, we deal with the weather. Is it warm outside? Is the wind strong? Is it cloudy or sunny? Is it raining or snowing? When we talk about the weather, we are referring to all of these conditions (Fig. 7.1). Fog, blizzards, thunderstorms and even sandstorms and hurricanes are also weather phenomena. Thus, weather is the state of the atmosphere and of the Earth's surface (including land and water) at a particular point in time. Weather is made up of meteorological components: atmospheric pressure systems, temperature and air humidity (near the Earth's surface), wind speed and direction, cloudiness and precipitation.

© The Author(s), under exclusive license to Springer Nature
Switzerland AG 2023
D. Livanov, *The Physics of Planet Earth and Its Natural Wonders*,
https://doi.org/10.1007/978-3-031-33426-9_7

Fig. 7.1 The vagaries of the weather: **a** snow in Moscow, **b** rainy weather in St. Petersburg, **c** sunny weather in Sevastopol (Russia), **d** a sandstorm in Khartoum (Sudan)

The term *climate* is used to refer to data that have been collected and averaged over a long period of time. Climate differs at different points on the Earth's surface; the principal reason for this difference is that the amount of sunlight that reaches it at different points is not the same. Therefore, the climate in the Sakha Republic of Russia does not resemble the climate in, for example, the suburbs of Moscow. Data about the climate in different locations give us an idea of what the Earth's climate is like in general.

It would seem that there is not a fundamental difference between the concepts of *weather* and *climate*. We know that weather is the state of the atmosphere at any given moment and given time and climate is the same thing, with the exception being that it refers to the entire Earth and lasts for an extended period of time. But, in actuality, the difference between these two terms is quite significant. If we take an interval average of, say, 10 years, changes will be evident in different ways on different parts of the Earth. This means, however, that the overall number of changes that occur over the entire surface of the Earth will be next to nothing. Let's say that in one particular area winter one year was especially warm, but the following year it was extremely cold. We would say that last year was warmer than this year, but this fact does not mean we can make a conclusion about climate change. If we average the data about the weather everywhere on the Earth, on a global scale we will see that the weather last winter was, on the contrary, colder than the current one. However, this still does not provide enough of a basis to ascertain that global warming is occurring. Climate change and the phenomena that impact upon the entire surface of our planet become apparent only after an average of 100 years or more. We can say, for example, that it was a few degrees warmer on the Earth in the eleventh and twelfth centuries than it is now, while in the sixteenth and seventeenth centuries the temperature was, in contrast, several degrees colder.

7.1   What Determines the Climate?

The ancient Greeks noticed that in certain places the climate is determined by the solar incidence angle at that specific location (in Greek the word *climate* means *angle*). This angle determines warming conditions; specifically, the amount of heat that the Sun gives off in a particular area. The key variable that affects the climate is the amount of energy supplied per unit area to the Earth in a 24-h period. Luminosity (i.e., illumination) per unit area is proportional to the square of the distance to the radiation source (i.e., the sun) and to the cosine of the angle between the surface normal and the sun's vector:

$$E = S \left(\frac{R_{E-S}}{r} \right)^2 \cos \alpha, \qquad (7.1)$$

where $S = 1.36 \times 10^3 \, \text{W/m}^2$ is the solar constant. This value is simply the intensity of illumination of the area located at the distance R_{E-S} from the Sun and its direct rays. But r is actual distance from the Earth to the Sun at a specific moment. Here angle α is the angle between the Sun's vector and the surface normal at a specific point on the Earth's surface.

It stands to reason that in this formula, distance depends on time (due to the ellipticity of the trajectory distance changes throughout the year, which causes an approximately 6% difference in illumination) and angle α (it changes throughout the course of a day and a year).

In order to determine the dependence of $\alpha(t)$, we will first find how angle γ between the Earth's rotation axis and the Sun's vector changes throughout the course of a year. Thereafter, in order to obtain the initial value of γ we will determine how the angle of the Sun's rays will change depending on the time of day and the geographic coordinates of a given area on the Earth's surface.

Calculating the Angle Between the Earth's Rotation Axis and the Sun's Vector We will denote the angle between the Earth's rotation axis and the sun's vector as γ.

We will roughly consider the angular velocity of the Earth as constant; in other words, we will ignore the ellipticity of the Earth's orbit. In this situation, within the time t the Earth will orbit the angle $\frac{2\pi t}{T}$, where T is a tropical year. We will denote the unit vector of the direction from the Earth to the Sun as $\mathbf{s}$, and the unit vector of the Earth's rotation axis as $\mathbf{m}$. We will incorporate the coordinates x, y and z as shown in Fig. 7.2. In these coordinates, the vector $\mathbf{s}$ has the components:

$$s_x = -\cos\frac{2\pi t}{T}; \; s_y = -\sin\frac{2\pi t}{T}; \; s_z = 0. \tag{7.2}$$

The components of vector $\mathbf{m}$ are:

$$m_x = \sin\varepsilon; \; m_y = 0; \; m_z = \cos\varepsilon. \tag{7.3}$$

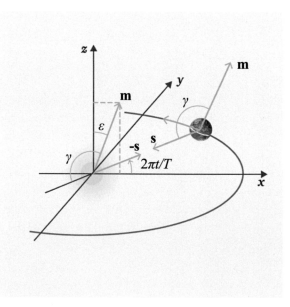

Fig. 7.2 The frame of reference for calculation of the angle between the Earth's rotation axis and the Sun's vector

It is possible to find the angle in question γ based on the fact that the cosine of the angle between the unit vectors is the sum total of the products of their projections:

$$\cos \gamma = (\mathbf{s} \times \mathbf{m}) = -\sin \varepsilon \times \cos \frac{2\pi t}{T}. \tag{7.4}$$

This formula determines the angle between the Earth's rotation axis and the Sun's vector depending on the time of year.

We will check the validity of this formula based on several situations that we know well.

1. The winter solstice: $t = 0$; $\cos \gamma = -\sin \varepsilon$; $\gamma = \varepsilon + \pi/2 = 113°$.
2. The spring and fall equinoxes: $t = T/4$ or $t = 3T/4$; $\cos \gamma = 0$; $\gamma = 90°$. Here the Earth's axis is perpendicular to the Sun's rays.
3. The summer solstice: $t = T/2$; $\cos \gamma = \sin \varepsilon$; $\gamma = 66°$.

The energy change in E that takes place throughout the year determines the characteristics of seasonal change, which is the primary factor that impacts upon the weather. In order to calculate this energy, one must average the cosine of the solar zenith angle

$$\cos \alpha = \cos \gamma \sin \varphi - \sin \gamma \cos \varphi \cos \omega t_{\text{lt}} \tag{7.5}$$

according to local time. This formula determines how this angle depends on the time of year, the time of day and the latitude of a point on the Earth's surface. In Fig. 7.3, corresponding references are shown. It is easiest to average a polar day because the second summand on the right side is zero. Then the value

$$Q = T_0 S \left(\frac{R_{\text{E−S}}}{r} \right)^2 \cos \gamma \sin \varphi, \tag{7.6}$$

will be equal to the amount of energy during the course of one day that reaches an area on the Earth's surface where there is midnight sun. T_0 is the length of a day, which is 86,400 s.

Dependency of the Solar Incidence Angle on the Time of Day and Geographic Coordinates φ and α We will denote the angle between the Sun's vector and the surface normal of the Earth as α. Let's analyze the solar incidence angle at a point on the Prime Meridian specifically at a certain latitude φ and longitude $\lambda = 0$ at the point of time t (Fig. 7.3). We will assume that the Earth revolves around the Sun with the constant angular velocity ω. The point of time t correlates to the Earth's rotation at the angle ωt with respect to the position that the Earth occupied at midnight Greenwich Mean Time. $\mathbf{s}$ is the unit vector of the Sun's

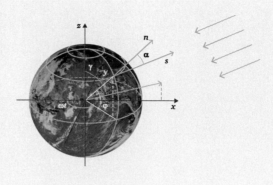

Fig. 7.3 Solar incidence angle at a point on the Prime Meridian

vector, which forms an angle γ with the axis z. As is shown in the illustration, its projections are equal to:

$$s_x = \sin \gamma; \; s_y = 0; \; s_z = \cos \gamma. \tag{7.7}$$

For the point on the Earth's surface that we have chosen, the unit vector of the normal line **n** has the following components:

$$n_x = -\cos \varphi \times \cos \omega t; \tag{7.8}$$

$$n_y = -\cos \varphi \times \sin \omega t; \tag{7.9}$$

$$n_z = \sin \varphi. \tag{7.10}$$

The scalar product of these vectors is the unknown cosine of the solar zenith angle:

$$\cos \alpha = \cos \gamma \times \sin \varphi - \sin \gamma \times \cos \varphi \times \cos \omega t. \tag{7.11}$$

Let's see how this formula works in several special situations.

1. Midnight: $t = 0$, therefore, $\cos \alpha = -\sin(\gamma - \varphi)$. The cosine turns out to be less than zero for almost all latitudes, which, from a physical standpoint, means that much of the Earth's surface is covered in darkness at midnight. $\cos \alpha = 0$ only pertains to a small section of the Earth's surface, which is where one finds a polar day. It is possible to find an equation for latitude above which a polar day occurs by labeling it as $\varphi_{\text{polar day}}$. Its endpoint is determined by the fact that the Sun's rays fall along a line tangent to the Earth at midnight: $\alpha = 90°$. Thus, the latitude $\varphi_{\text{polar day}}$ is determined by the equation: $\sin(\gamma - \varphi_{\text{polar day}}) = 0$. Two values for $\varphi_{\text{polar day}}$ satisfy this equation: $\varphi_{\text{polar day}} = \gamma$ and $\varphi_{\text{polar day}} = \gamma - \pi$.

The first value corresponds to the endpoint for a polar day in the Northern Hemisphere between the spring and fall equinox, while the second value corresponds to the endpoint of a polar day in the Southern Hemisphere between the fall and spring equinox.

2. Noon: $\omega t = \pi$. We get: $\cos\alpha = \sin(\gamma + \varphi)$. The solution to this equation is $\alpha = \gamma + \varphi - \frac{\pi}{2}$. In order for a polar night to begin, a negative cosine value is required. The corresponding latitude $\varphi_{\text{polar night}}$ is determined by the equation $\sin(\gamma + \varphi_{\text{polar night}} = 0)$, the solution to which is $\varphi_{\text{polar night}} = -\gamma$ and $\varphi_{\text{polar night}} = \pi - \gamma$. The first solution corresponds to the Southern Hemisphere, while the second one corresponds to the Northern Hemisphere.

3. We will find the duration of daylight as a function of latitude. Sunrise and sunset correspond to the direction of the Sun's rays along a line tangent to the Earth $\cos\alpha = 0$. Since night lasts twice as long as morning, we get:

$$t_{\text{night}} = \frac{P_0}{\pi}\arccos\left(\operatorname{tg}\varphi \times \operatorname{ctg}\gamma\right). \tag{7.12}$$

If $t_{\text{night}} \leq P_0$ (the duration of a 24-h period), it is easy to find the length of a day: $t_{\text{day}} = P_0 - t_{\text{night}}$. The variation $t_{\text{night}} > P_0$ corresponds to a polar night. A polar day corresponds to an alternative variation when $\operatorname{tg}\varphi \times \operatorname{ctg}\gamma > 1$.

4. Fall and spring equinox: $\gamma = \frac{\pi}{2}$. In this situation, $\cos\alpha = -\cos\varphi \times \cos\omega t$. All latitudes have sunrise at 6:00 a.m. and sunset at 6:00 p.m. The length of a 12-h day is equal to the length of a night.

5. The Equator: $\omega = 0$, $\cos\alpha = -\sin\gamma \times \cos\omega t$. At the Equator, a day lasts 12 h regardless of the time of year.

When dealing with an arbitrary latitude, it is more difficult to average cos α over time because illumination at night can take positive values; in this instance, numerical integration will help us. In Fig. 7.4, we see an example of the calculated dependency of average illumination as determined by the time of the year at 60° north and south of the Equator.

In Fig. 7.5, one sees how temporal variation of the average annual temperature, which has been experimentally measured, appears at different latitudes.

We see that the observed results are similar to the results derived from theoretical calculations. There are, however, two factors that complicate this fairly simple model for calculating solar energy distribution on the Earth. The first one is the redistribution of energy on the Earth's surface due to heat transfer, which results in an extremely low temperature in places near the North and South Poles. The second factor is heat retention by the atmosphere, dry land and the ocean. The word *retention* here applies to the special properties of a substance, which cause its thermal state to produce a delayed time response when changes in external conditions—first and foremost in solar illumination—take place. In other words, after changes in illumination

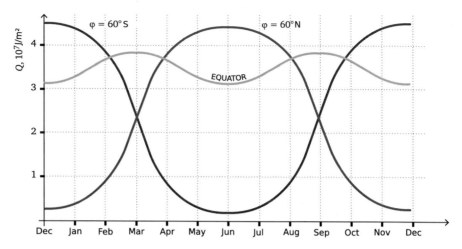

Fig. 7.4 Changes in diurnal light energy at the Equator during the course of the year at 60° north latitude and 60° south latitude

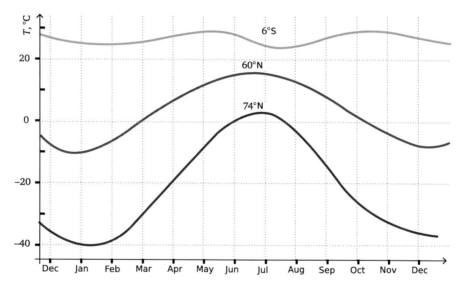

Fig. 7.5 Changes in the average annual temperature at different latitudes

occur, an increase in temperature does not immediately follow, but rather is observed over a period of time. This estimated time provides a valid explanation of the approximate monthly lag between the minimum and maximum average temperature and the minimum and maximum solar illumination.

Time Lag Between a Temperature Change in the World Ocean and a Change in Luminous Intensity As we know, currents mix the top layers of water in the ocean down to the depth of $h \approx 100$ m (328.084 ft). We can estimate the total heat capacity of this actively intermixing layer on the ocean's surface as follows:

$$C \approx 0.7 \times 4\pi R_E^2 h \rho_{water} c_{water} = 1.7 \times 10^{23} \text{ J/K}, \qquad (7.13)$$

where the coefficient 0.7 reflects the fact that the ocean occupies about 70% of the Earth's surface and c_{water} is the heat capacity of water. Keeping in mind that the Sun's rays illuminate the Earth's surface with an intensity of $\pi R_E^2 S = 1.7 \times 10^{17}$ W, we will find the time needed to heat the ocean to $\Delta T = 3\,^\circ\text{C}\,(37\,^\circ\text{F})$:

$$\Delta t = \frac{C \Delta T}{\pi R_E^2 S} \approx 3 \times 10^6 \text{ s} \approx 1 \text{ month.} \qquad (7.14)$$

This estimate corresponds to an approximately one-month lag in a maximum and minimum temperature change with regard to a minimum and maximum change in luminous intensity as shown in Figs. 7.4 and 7.5.

To illustrate the difference between weather and climate, let's consider a situation that is well known to hikers. If a hiker walks across a mountain into the wind, they will immediately notice a significant difference in weather on the leeward and windward sides of the mountain. If while climbing up the mountain from the leeward side this individual gets caught in heavy cloud coverage and fog or, even worse, a downpour or a snowstorm, after reaching the summit, they will be met by sunny, warm weather on the windward side.

The sun shines the same on both sides of the mountain, which is why the climate is absolutely the same there (although it is incorrect to talk about the climate when referring to one or two specific points). The weather, on the other hand, will be completely different and will depend on which side of the mountain the wind is blowing.

The qualitative explanation for this effect is as follows. When wind comes in contact with a large mountain, air masses begin to intensively rise up. When air moves to a low-pressure area, it expands. This process can be considered adiabatic, which means that the expansion of air will be followed by a decrease in internal energy or, in other words, cooling will occur. Since the pressure of saturated water vapor decreases with a drop in temperature, some of the moisture present in the air will condense and fall in the form of mist or rain and snow.

This dry air passes over the mountain's peak and begins to descend. Because of its low level of humidity, when it descends, the temperature of this air mass rises faster than it would have risen for a mass that had the original amount of humidity. Hence, if altitude is the same, then the air temperature on the windward side of the mountain will be higher and the humidity will be lower than on the leeward side.

Weather Conditions on Different Sides of a Mountain Let's assume that air with the mass M blows over a mountain. This body of air is on the leeward side of the mountain and has the volume V_1, the temperature T_1 and the pressure p_1. After reaching the peak and descending down the windward side, it will be characterized by the volume V_2, the temperature T_2 and the pressure p_2 (Fig. 7.6). Since the internal energy of one mole of diatomic gas is $\frac{5}{2}RT$ ($R = 8.3 \times 10^3 \frac{J}{\text{kmole K}}$), the change in the internal energy of the emitted air mass will be equal to:

$$\Delta U = \frac{5}{2} \frac{M}{\mu_{\text{air}}} R(T_2 - T_1). \tag{7.15}$$

But what exactly is the function of a body of air that rises and then falls? By doing negative work $-p_1 V_1$ on the leeward side of the mountain, it displaced the air in volume V_1. On the other hand, the descending air did positive work $p_2 V_2$ on the windward side of

Fig. 7.6 Weather conditions on opposite sides of a mountain

the mountain and pushed out the air in volume V_2. Thus, this entire operation is equal to

$$A = p_2 V_2 - p_1 V_1. \tag{7.16}$$

Using the ideal gas law $pV = \frac{M}{\mu} RT$, we get

$$A = \frac{M}{\mu_{\text{air}}} R(T_2 - T_1). \tag{7.17}$$

We will assume that when air rises, it loses practically all of its moisture that fell in the form of rain or snow. This assumption is realistic if a mountain is very tall. We will denote a water mass that fell in the form of precipitation as Δm. The amount of heat $Q = \lambda \Delta m$ was then released during condensation.

We will assume that the air humidity from the leeward side is such that when the air pressure is p_1, the vapor pressure is equal to p. We then have the following set of equations:

$$\begin{cases} pV_1 = \frac{\Delta m}{\mu_{\text{water}}} RT_1, \\ p_1 V_1 = \frac{M}{\mu_{\text{air}}} RT_1, \end{cases} \tag{7.18}$$

which when we solve, we get

$$\Delta m = M \frac{\mu_{\text{water}}}{\mu_{\text{air}}} \frac{p}{p_1} \tag{7.19}$$

We will find the amount of heat that is emitted:

$$Q = \lambda M \frac{\mu_{\text{water}}}{\mu_{\text{air}}} \frac{p}{p_1} \tag{7.20}$$

By virtue of the law of conservation of energy,

$$Q = \Delta U + A. \tag{7.21}$$

If we substitute the formulas here that we found, we can find the temperature difference on the mountain slopes:

$$T_2 - T_1 = \frac{2}{7} \frac{\lambda \mu_{\text{water}}}{R} \frac{p}{p_1}. \tag{7.22}$$

Let's assume that $p_1 = 10^5$ Pa. When the air humidity is 50% and the temperature is 18 °C (64 °F), we then have $p \approx 0.01 \times 10^5$ Pa. Finally, we will find: $T_2 - T_1 = 15\,°C\,(59\,°F)$. We learn that the temperature difference on the leeward and windward sides of the mountain is quite substantial.

The side of the mountain where the temperature is higher obviously does not have any bearing on climate conditions because it does not affect the Earth's average temperature in any way. However, for a shepherd grazing cattle on the slopes of this mountain, let's say, it is very important whether he will get soaked in the rain on the windward side or bask in the sun's warmth on the leeward side. This is exactly how the weather works: it is determined not as much by climate conditions as it is by local factors at a particular moment in time.

When we were describing the wind, we were certain that on average an air mass travels around the globe in about one week. This is exactly the characteristic time that weather changes occur on the Earth. These changes are caused when short-term variations of physical quantities, which determine the weather, differ from their average estimates. The climate is another matter altogether. It can also change over time, but these changes, if they do indeed occur, happen much more slowly.

7.2   Clouds

Clouds are a source of precipitation on the Earth, which is the reason they are important to the weather that develops everywhere. But clouds also play a part the in Earth's entire climate. Indeed, their albedo is quite high and, as we have already established, solar energy, which is the primary determinant of the climate, absorbed by the Earth strongly depends on clouds' albedo. For this reason, without studying the structure of clouds and their characteristics, it is impossible to predict the weather and project trends in climate change.

Many people think that clouds are made up of water vapor. This is a very common misconception because water vapor is invisible. We can check this if we look at the very edge of a boiling tea kettle's spout. Although the steam is invisible, when it cools down and turns into droplets, at a certain distance away from the tea kettle we can see a whitish cloud (Fig. 7.7). This is most definitely an honest-to-goodness little cloud, albeit a very short-lived one because as soon as the tea kettle is removed from the stove, the cloud disappears.

The movement of clouds in the sky raises the following questions: Why don't "real" clouds in the sky fall down onto the Earth? Why do some clouds cause a downpour and others do not? And why are certain clouds typically white, while others such as storm clouds are usually a darker shade of gray?

Let's start with the last question, since the answer to it is the easiest. Droplets in a cloud are usually longer than the wavelength of visible light.

Fig. 7.7 A cloud over the spout of a boiling tea kettle

In this case, an incident beam is reflected off a droplet's outer surface. Nevertheless, its light remains white, which is the natural mixture of colors of the solar spectrum. Storm clouds look dark gray to us because they let very little sunlight pass through them. In such clouds as these, light is either reflected upward or absorbed by droplets. The term *storm cloud* is not used by scientists; rather, they only speak about clouds. The difference between white, gray and nearly black clouds is determined by so-called *optical depth*, which is the number of droplets that are directed toward a beam of sunlight as it passes through a cloud. If there are only a few droplets, they randomly change the direction of sunlight but barely affect its intensity. Light becomes diffused just as it does from a frosted light bulb. If there are many droplets in a cloud, a significant amount of light is absorbed and the cloud is dark.

What determines the location of the base and the top of a cloud? When air rises upward, it expands adiabatically, but at the same time its temperature drops. When the temperature reaches the dew point, which is the temperature at which water vapor becomes saturated, vapor condenses into small droplets. This is exactly what determines where the base of a cloud is. When this base reaches the ground, we say that fog is "hugging" the ground.

Fog is a suspension of drops in the air directly above the surface of land or water. The average size of droplets that cause fog to form is about $10\,\mu m = 10^{-5}$ m. As we have already pointed out, fog is, in fact, a cloud with a base that extends to the Earth's surface. The main condition that must be met in order for fog to form is that at a given temperature, there must be more partial water vapor pressure in the air than saturated vapor pressure.

Let's recall that pressure is partial when there is one element in a gas mixture; the total pressure of a gas mixture is equal to the sum total of the partial pressure of elements.

If this is the case, then droplets will emerge in an air mass, and, furthermore, any type of particles suspended in the air—for example, dust particles—may act as a nucleation center. Fog may form due to two possible key scenarios.

The first scenario involves local air cooling. Let's suppose that the air temperature drops when the pressure is stable. Let's assume that point A is characterized by the initial pressure p_1 and by the temperature T_1 of water vapor (Fig. 7.8). For this type of pressure, the temperature T_2 is the dew point. If the temperature drops to the value $T_3 < T_2$, a state of supersaturated vapor will result, which, when it breaks down, will cause an emission to occur on the nucleation center of each droplet.

Air temperature may rapidly drop, when, for example, hot air moves from land to sea. In this situation, when air warms the cold sea, it cools down and fog forms over the water. Another situation is also possible in winter when warm air that has been over a body of water moves to a cold, snow-covered shore. Evening mist (Fig. 7.9) forms in summer when the ground that was warmed by the sun's rays quickly cools down after sunset and the temperature of the subsurface air layers drops with it.

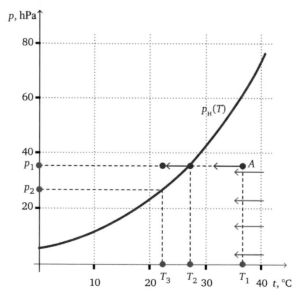

Fig. 7.8 Diagram illustrating the components of fog formation when pressure is kept constant and the temperature drops (i.e., *advection fog*)

Fig. 7.9 Evening mist

The second scenario that explains the formation of fog plays out when there is a local pressure increase of water vapor due to active evaporation, for example, in the morning from the surface of a water reservoir. We will denote the reference state of the vapor with point B as shown in Fig. 7.10. If vapor pressure rises at a constant temperature, then provided that $p > p_2$, mist will begin to fall. This type of situation is common in the morning over the surface of a water reservoir. Since the air over the water cools down faster overnight than the water, the water's surface begins to actively evaporate, which increases water vapor pressure.

But the escapades of drops in clouds do not stop there. As you know, when water vapor condenses, a good deal of heat is released, which warms up the air. It becomes slightly warmer than the surrounding air, expands and rises up because of the buoyant force. Then it adiabatically expands again, cools down to the temperature around it and stops. This is where the cloud's top is located.

Temperature Drop in a Cloud During an Adiabatic Expansion The fundamental process that is responsible for the development of clouds is an adiabatic expansion of an air mass as it is rising.

Let's consider the expansion of a diatomic gas that has the mass M. During this process, the gas does work against the pressure force from the air around it. Let's assume that while expanding, the gas receives the amount of heat Q from outside, which it spends on doing the work A and on changing its internal energy ΔU:

$$\Delta U = Q - A. \tag{7.23}$$

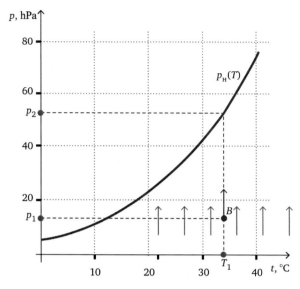

Fig. 7.10 Diagram illustrating the components of fog formation when pressure increases and the temperature remains constant (i.e., *steam fog*)

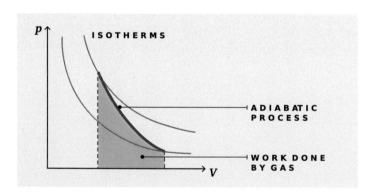

Fig. 7.11 Isoterms and adiabat in the p–V plane

A change in the gas's internal energy causes a change in its temperature. Let's assume that T_1 and T_2 are the initial and the final temperatures of the gas, respectively.

A change in the internal temperature will be equal to:

$$\Delta U = \frac{5}{2} \frac{M R (T_2 - T_1)}{\mu}. \tag{7.24}$$

When a cloud forms, this air mass quickly rises up and usually there is not enough time for a heat exchange to take place. Hence, $Q = 0$. In this situation, $\Delta U = -A$ and $\Delta U < 0$;

thus, $T_2 < T_1$, i.e., when gas expands, it cools down. Cooling results because of the work that is done only by the internal energy of the gas without a heat exchange with the external environment. When there is a temperature change during adiabatic cooling of a gas, we get:

$$T_2 - T_1 = \frac{2}{5} \frac{\mu A}{MR} \qquad (7.25)$$

We will find the work of the gas during the adiabatic process. After differentiating the Clapeyron–Mendeleev equation $pV = \frac{M}{\mu}RT$, substituting the formula for internal energy in terms of the temperature for diatomic gas $dU = \frac{5}{2}\frac{M}{\mu}RT$ into it and using the formula $dU = -dV$, we get:

$$\frac{7}{5}pdV = -Vdp. \qquad (7.26)$$

The solution to this equation is an adiabatic equation at the coordinates $\{p, V\}$:

$$pV^{\frac{7}{5}} = \text{const.} \qquad (7.27)$$

In diagram form, an adiabatic curve at these coordinates looks exactly as it is shown in Fig. 7.11.

Similarly, we get

$$TV^{\frac{2}{5}} = \text{const}, \ Tp^{-\frac{2}{7}} = \text{const.} \qquad (7.28)$$

Let's assume that p_1 and p_2 is the initial and the final gas pressure. When we express the temperature in terms of pressure using the appropriate adiabatic equation, we get a formula for the work that the adiabatically expanding gas performs:

$$A = \frac{5}{2}\frac{M}{\mu}RT_1\left(1 - \left[\frac{p_2}{p_1}\right]^{\frac{2}{7}}\right). \qquad (7.29)$$

For example, if the air temperature near the Earth's surface is 27 °C (80.6 °F), the air rises up 1 km (0.62 mi) and the air pressure at this height is 90% of the pressure near the Earth's surface, we can find that $T_2 - T_1 = 9$ K. Therefore, when adiabatically expanding air rises up 1 km (0.62 mi), it cools down 9°.

We will once again describe the phenomenon of cloud formation. When air rises, it cools down. At a certain height, it will cool down so much that water vapor will begin to condense. This is the height H and it will determine the cloud's base (Fig. 7.12). Air coming from below will pass through this base and because of additional condensation, the cloud will increase in height. Finally, at a certain height the air will stop rising. This will determine the location of the cloud's top. As we know, a certain amount of heat is released during vapor condensation. Therefore, the temperature of rising, moist air will not drop with altitude as quickly as the temperature of surrounding, stagnant dry air.

What happens next? Since the temperature of the rising air is higher than the ambient temperature, the cloud continues to increase in height. Since the rising air becomes drier and drier, at a certain altitude its temperature equals that of the ambient temperature. At that time, the cloud stops developing vertically. This is where the cloud's top will be. The cool, dry air will then begin to spread out in all directions and form fleecy white clouds, which are so distinctive among all cloud types (Fig. 7.13).

Types of Clouds All clouds can be divided into several types (Fig. 7.14).

Nimbostratus, stratocumulus and stratus clouds form at an altitude that does not exceed 2 km (6561.68 ft). These are low-level clouds. Stratus clouds form either when a stagnant air mass cools down (e.g., at night) or when warm air moves over the Earth's cool surface.

Altocumulus and altostratus are mid-level cloud types whose height does not exceed 2–7 km (6561.68–22,965.9 ft).

Cirrus, cirrocumulus and cirrostratus are high-level clouds that are found at an altitude of 7–15 km (22,965.9–49,212.6 ft). Cirrus clouds form at a high altitude of up to 15 km

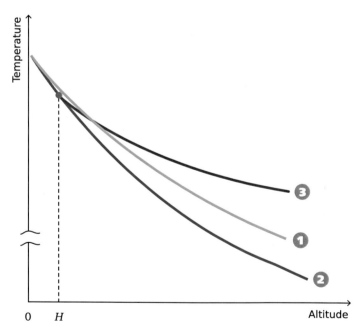

Fig. 7.12 Air's dependence on height: (1) for air without clouds; (2) for expanding dry air in clouds; (3) for expanding moist air in clouds

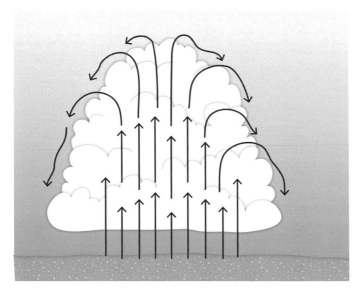

Fig. 7.13 Diagram of air movement when a cumulus cloud forms

(49,212.6 ft) in strong air currents and are made up of small crystals of frozen water, that is, ice.

In addition, there are clouds that develop vertically and stretch down several levels from the high altitudes. These are cumulus and cumulonimbus clouds.

Cumulus clouds form when there is convection of moisture-filled air. They often look like puffy balls of cotton. If convection is strong, a rain cloud forms, which usually has a top that is 7–10 km (22,965.9–32,808.4 ft) above the Earth's surface.

It is quite common to see clouds "capping" the tops of mountain peaks (Fig. 7.15). This happens because convection causes warm air to rise up the mountain, which is followed by moisture condensation when the temperature drops.

At high altitudes strong winds try to carry these clouds away, but they stubbornly hold tight to the mountain's peak (Fig. 7.16).

Basic formulas explain the phenomena behind clouds' formation and allow us to calculate their dimensions, but their specific shape depends on chance factors and is therefore unpredictable. It is not a coincidence that we can find thousands, if not millions, of pictures on the Internet of beautiful clouds with the most unusual shapes.

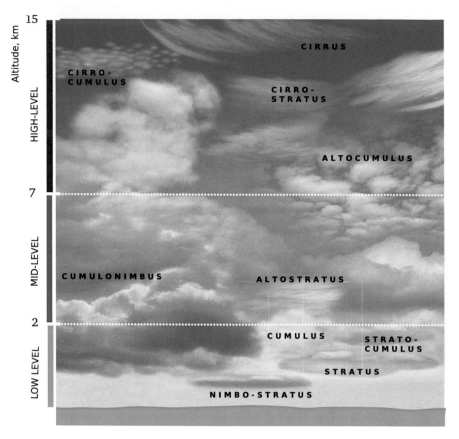

Fig. 7.14 Types of clouds

7.3   Drops in the Air

As we have already learned, clouds are made up of very small droplets that hang in the air. It would seem that there isn't anything of great interest to say about them. Based on images of drops in cartoons and children's books, we imagine that they bear somewhat of a resemblance to tadpoles with a head and tail, which is actually quite far from the truth. That is what drops look like when they are hanging from a rooftop, but when they fall or "fly" in clouds, drops look quite different. In this section, we will analyze both the physics of the levitation of drops in the air and their shape.

The ability of drops to stay in the air depends on their size.

If the size of a drop of water is significantly less than 1 μm, it does not fall down but rather makes fast and chaotic movements. It is specifically these drops that make up the white, fluffy clouds that never produce rain.

Fig. 7.15 A cloud over Mt. Fuji

Fig. 7.16 A cloud over the top of Mt. Everest. A view from the top of Kala Patthar

Size of a Drop in a Cloud Let's try to calculate the maximum size of a drop of water that is unlikely to fall from a cloud as a raindrop. We will assume that the drop is round because surface tension tries to configure it into a shape that has a minimum surface area for the given volume. The smallest droplets that have a diameter of less than 1 μm (10^{-6} m) cannot even fall to the Earth's surface. The reason for this is because after these droplets enter into the environment of air molecules (Fig. 7.17a), they move in a chaotic manner once they

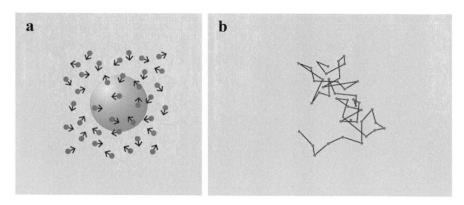

Fig. 7.17 Water droplet in the environment of air moleclules and example of its chaotic path

have been exposed to impingement. As we know, air molecules have a mass of approximately $m \approx 5 \times 10^{-26}$ kg and move at the speed $v \approx \sqrt{\frac{kT}{m}} = 300$ m/s (984.252 ft/s).

Since kinetic energy is transferred from air molecules to drops during impingement, the amount of kinetic energy of a drop of water affected by Brownian motion with the mass $m = \rho_{water}V$ is $E = \frac{\rho_{water}Vv^2}{2}$. We will make this energy equal to the average thermal energy of air molecules $U = kT$. Let's assume that the radius of a drop is equal to R. Then its volume is $V = \frac{4}{3}\pi R^3$. Based on this information, we can find the velocity of the Brownian motion of the drop:

$$v = \sqrt{\frac{3}{2\pi}\frac{kT}{\rho_{water}R^3}}. \tag{7.30}$$

Now let's imagine that we are dealing with a larger drop, which starts to move downward because of the force of gravity acting on it. Both the downward force of gravity, which is equal to $F_g = \rho_{water}gV$, and the upward buoyant force from the atmosphere as per Archimedes' principle, which is equal to $F_A = \rho_{air}gV$, are acting on the falling drop of water. Since $\rho_{water} = 10^3$ kg/m^3 and $\rho_{air} = 1.3$ kg/m^3 (at 4 °C; 39.2 °F), the buoyant force is negligible compared to the force of gravity. If only two forces were acting on the drop, it would have steadily accelerated downward. As it is, a third force—the viscous force (i.e., Stokes' law)—is actually acting on it. It is clear that the higher the medium viscosity, the stronger this force will be. Air viscosity is equal to $\eta = 1.8 \times 10^{-5} \frac{kg}{m\,s}$. In order to deduce the degree of force ($H = \frac{kg\,m}{s^2}$), the viscosity must be multiplied by the drop's rate of fall and its radius.

If we do an exact calculation of a spherical body, we get the numerical coefficient 6π. Owing to this, air resistance is equal to $F_{AR} = 6\pi\eta vR$. If we ignore the buoyant force and make the force of gravity equal to air resistance, we can find a drop's fall rate:

$$v = \frac{2}{9}\frac{\rho_{water}gR^2}{\eta}. \tag{7.31}$$

Now we can determine the radius of the drop whereby the velocity of Brownian motion is equal to the fall rate due to gravity. This radius is equal (up to the numerical coefficient) to:

$$R_0 \approx \left(\frac{kT\eta^2}{\rho_{water}^3 g^2} \right)^{\frac{1}{7}} \approx 10^{-6}\,\text{m}. \tag{7.32}$$

It is important to emphasize that the force with which air acts on a falling drop of water moves at a linear speed if it is relatively low (the corresponding Reynolds number is $Re \ll 1$). When drops move faster, turbulent air, which flows around them and is caused by a countercurrent, begins to act on these drops. Stokes' law then turns into drag force, which is proportional to the squared velocity.

If the size of the drop is about 1 μm (10^{-6} m), it starts to move toward the Earth's surface. These drops' fall rate is low—about 1 mm (0.039 in)/s.

In actuality, motion in viscous environments (i.e., liquids and gases) actually occurs in a much more complicated manner. Not all motion is associated with the plane-parallel flow of a liquid or gas. Motion that occurs at a slow speed whereby liquid molecules move in a parallel direction is called *laminar flow*. When velocity increases beyond a certain level, laminar flow is replaced by so-called *turbulent flow*. When this occurs, molecules move along trajectories with vortices that flow in random directions. During turbulent motion, the air currents that flow around drops of water actively mix. The typical value of the size of a drop at which laminar flow changes to turbulent flow is $R_1 \approx 10^{-4}$ m. The drag force that acts on a drop is equal to $F_{turb} \approx \rho_{air} v^2 R^2$. After making the drag force equal to the force of gravity, we find that the velocity of these drops is $v \approx \sqrt{\frac{\rho_{water} g R}{\rho_{air}}}$. Thus, they move at approximately 1.5 m (59.055 in)/s.

Therefore, let us assume that a drop of water that is about the size of $R_0 = 10^{-6}$ m will start to move downward due to gravity. As it moves it will absorb smaller droplets; thus, the larger the drop's size, the greater the number of droplets that it will absorb.

Eventually, at a certain critical point, there will not be anything to keep this drop from beginning to fall. We obviously know that when there is a large number of drops, it starts to rain.

What happens when drops of water continue to grow in size? The answer is that they are no longer round. This happens because the force of surface tension is too weak for them to retrain their spherical shape and they begin to flatten out. The value of water's surface tension is: $\sigma_{water} = 7.2 \times 10^{-2}\,\text{kg/s}^2$, and the respective force is equal to $F_{ST} = \sigma_{water} R$. We will determine the radius of the drop R_2 whereby the value of the surface tension force F_{ST} and the air resistance in turbulent conditions F_{turb} will be the same. We get:

$$R_2 \approx \sqrt{\frac{\sigma_{water}}{g\rho_{water}}} = 3 \times 10^{-3} \text{ m.} \tag{7.33}$$

The Minimum Height of a Rain Cloud Let's estimate how high a cloud needs to be in order for it to produce rain. We will observe falling drops for a short period of time dt. Within that time, drops will fall the distance $dh = vdt$. The absorption cross-section of little drops by larger ones is determined by the area $S = \pi R^2$. While we are observing these drops, they will collect all of the available moisture in the amount:

$$V = \pi R^2 vdt. \tag{7.34}$$

The moisture mass that was collected is determined by multiplying the aforementioned amount by the cloud's liquid water content w (this is the mass of water per unit volume of air, which is measured in g/m^3). The drops' mass increases to the amount:

$$dm = \pi R^2 wvdt. \tag{7.35}$$

We will substitute the drops' flow rate in laminar flow into this formula and we then get:

$$dm = \frac{2\pi}{9} \frac{w\rho_{water}gR^4}{\eta}dt. \tag{7.36}$$

Based on the relationship between the moisture mass and the drops' radius $dm = 4\pi\rho_{water}R^2dR$, we obtain an equation that describes the drops' increase in size or, in other words, joins the drops' radius and time:

$$\frac{dR}{R^2} = \frac{wg}{18\eta}dt. \tag{7.37}$$

If the drops increase in size from R_0 to $R \gg R_0$ within the time τ, we get:

$$\tau = \frac{18\eta}{wgR_0} \approx 6\,\text{h}/w. \tag{7.38}$$

Now we will estimate how much the drops increase when they move down. We have:

$$\frac{\Delta R}{\Delta h} = \frac{w}{4\rho_{water}}. \tag{7.39}$$

The drops must retain quite a large size—approximately R_2—so that they do not evaporate when falling. Therefore, the height of a cloud, which is the distance from its top to its base, that can produce rain is equal to:

$$\Delta h = \frac{4\rho_{water}}{w} R_2 \approx 4 \, km \, (13,123.4 \, ft).$$ (7.40)

Calculating the Shape of a Drop of Water We will estimate to what degree a drop deviates from its spherical shape and focus on a central column in it (Fig. 7.18). If this drop moves at a steady pace, the force of gravity that acts on it is balanced by surface tension. Therefore, the radii of curvature of the drop's surface at points A and B should not be the same. Moreover, since the force of gravity is directed downward, the drop's curvature at the bottom of the surface (A) should be greater than the curvature at the top (B).

We get:

$$\rho_{water} gh = \frac{2\sigma_{water}}{R_A} - \frac{2\sigma_{water}}{R_B}$$ (7.41)

We will compare the hydrostatic pressure (i.e., the left side) and the pressure from a curvature difference (i.e., the right side) of this equation when drops have various sizes. First, we will analyze a drop we assume is very small, i.e., $R = 10^{-6}$ m. In this case,

$$\rho_{water} gh \approx 2 \times 10^{-2} \, Pa, \text{ but } \frac{2\sigma_{water}}{R} \approx 10^5 \, Pa.$$ (7.42)

The difference in pressure is colossal and a drop such as this is absolutely spherical.

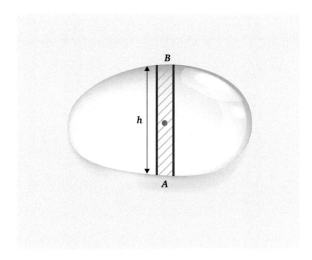

Fig. 7.18 Form of falling water droplet

Now let's analyze a drop with the diameter $R = 4 \times 10^{-3}$ m $= 4$ mm (1.73228 in). Here $\rho_{water}gh \approx 40$ Pa, but $\frac{2\sigma_{water}}{R} \approx 30$ Pa. Hence, the pressure difference is comparable and we get a significant deviation from sphericity. We will denote the difference in the radii of curvature at points A and B as $\Delta R = R_B - R_A$. Since $R_A + R_B = h = 4$ mm (1.73228 in), we get:

$$\Delta R \approx h \left(\sqrt{\left(\frac{\sigma_{water}}{\rho_{water}ghR} \right)^2 + 1} - \frac{\sigma_{water}}{\rho_{water}ghR} \right) \approx 1 \text{ mm } (0.0393701 \text{ in}). \qquad (7.43)$$

Thus, capillary pressure is comparable to hydrostatic pressure and a significant deviation from sphericity, which is approximately a quarter of the drop's size, results. The drop is actually highly deformed.

Drops that are approximately R_2 and greater are flattened by oncoming air flows. A drop's shape becomes irregular and changes as it moves down. When the drop is about 1 cm (0.393701 in), it eventually splits in half.

Figure 7.19 shows raindrops of different diameters that fall at different speeds.

The shape of a falling drop is determined by the fact that if it moves fast enough, air does not have time to flow around it, which results in the formation of an eddy current (i.e., turbulence). An area of low pressure forms on the top of the drop, while high pressure forms on the bottom of it (Fig. 7.20). Pictures of falling drops really do show that they are flattened out on the underside.

If a drop is falling at a high speed and the corresponding pressure difference is significant (i.e., higher than the hydrostatic pressure $\rho_{water}gh$), then

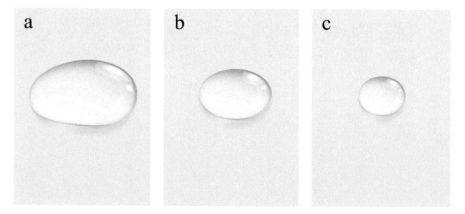

Fig. 7.19 Shape of drops of different sizes that fall at different speeds: **a** 6 mm (0.23622 in), speed—8.8 m/s (28.87 ft/s); **b** 5 mm (0.19685 in), speed—8.3 m/s (27.23 ft/s); **c** 3 mm (0.11811 in), speed—6.8 m/s (22.31 ft/s)

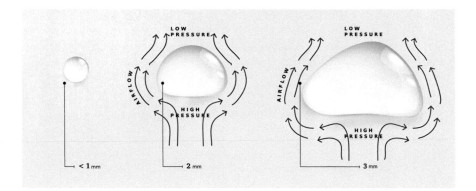

Fig. 7.20 The direction of air currents that flow around falling drops of various sizes

because the top and the bottom of the drop's surface $\frac{2\sigma_{\text{water}}}{R_A} - \frac{2\sigma_{\text{water}}}{R_B}$ are curved differently, the pressure difference changes its sign and becomes negative. In this situation, $R_B < R_A$, which is illustrated by the falling drops (Fig. 7.21).

When large drops of water fall, they cannot maintain their shape; fast-flowing air currents cause them to break up into little droplets. Slow motion video (Fig. 7.22) has captured images showing that first large drops become flattened out and turn into something that resembles flying flatbread. Then they take the shape of a canopy or parachute. After that, their top layer bursts into many small splashes of water.

Now you know that rain is a much more complicated phenomenon than it may seem.

Drops of Water on Leaves After it has rained, we can often see drops of water on plant leaves (Fig. 7.23). Some of these drops are almost perfectly round, while others are slightly flattened out.

We can see that the larger the drop, the larger and more shapeless it becomes, thus turning into something that resembles a thick crêpe. The surface tension of the liquid and the curvature of the drop's surface create pressure that attempts to make the drop round:

$$p_\kappa = \frac{2\sigma_{\text{water}}}{R}. \tag{7.44}$$

The force of gravity attempts to flatten out the drop. The pressure it creates—p_g—is equal to the ratio of the force of gravity mg to the surface contact area of water and the solid surface on which the drop lies. We will assume that this area is approximately the same as the square of the drop's

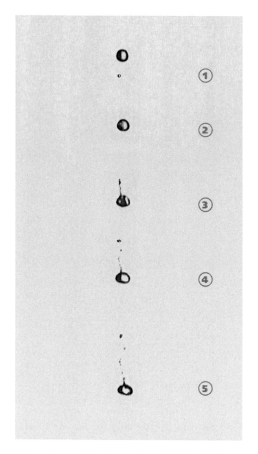

Fig. 7.21 Changes in the shape of a large drop in free fall

radius. Then

$$p_g = \frac{4\pi}{3} g \rho_{water} R. \tag{7.45}$$

In order for the drop to remain almost round, the required condition is that $p_g \ll p_\kappa$.

Thus, we learn that only small drops of water with a radius of

$$R \ll \sqrt{\frac{\sigma_{water}}{2g\rho_{water}}} \approx 2\,\text{mm}\,(0.0787402\,\text{in}) \tag{7.46}$$

will be round.

This estimate is true in the case of water that does not thoroughly moisten the surface of leaves.

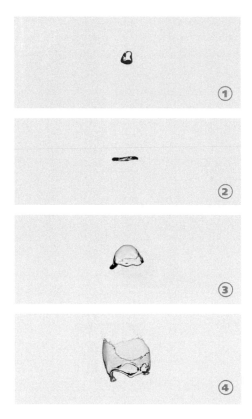

Fig. 7.22   Changes in the shape of a small drop in free fall

Fig. 7.23   Water drops on plant leaves after a rain

Drops of Water on Glass Everyone has watched drops of water sliding down a window pane when it was raining. When drops land on glass, they stick to it and then either stay there or start to slide down the glass (Fig. 7.24). When raindrops run down it, they usually leave behind a water tail. Sometimes drops that have stuck to glass get "caught" by larger ones as they run down a surface. In this case, drops blend together with others, which causes a curvature of their trajectory.

What are the physical reasons why raindrops on window panes do this? It is fairly obvious that the force of gravity is what makes drops slide down glass. But only quite large drops run down glass; smaller ones seem to be glued to it. What holds them in place? The force of interaction between the water molecules and the glass molecules does. This mutual effect is quantitively caused by energy that must be expended in order to separate the water from the glass when there is a unit area of contact. We will designate this value as $\sigma_{\text{water–glass}}$. When a drop is separated from a glass surface, two free surfaces form: a water surface with the energy σ_{water} and a glass surface with the energy σ_{glass}. The energy required to separate a drop of water from glass on a unit surface of contact is $\Delta\sigma = \sigma_{\text{water–glass}} - \sigma_{\text{water}} - \sigma_{\text{glass}}$. When the radius of the drop is R, the force that keeps the drop from running down it will be equal to $F_{\text{sep}} = 2R\Delta\sigma$. We will consider the drop to be a hemisphere. It will start moving if the force of gravity $F_{\text{g}} = mg = \frac{2}{3}\pi R^3 \rho_{\text{water}} g$ is greater than

Fig. 7.24 Water drops on glass

F_{sep}. We understand that the drop will slide down glass if its size is

$$R > \sqrt{\frac{\Delta\sigma}{\rho_{water}g}} \qquad (7.47)$$

If $\Delta\sigma > 2\sigma_{water}$, then a downward-moving drop will leave behind a water tail because it is energetically favorable for it to slide down a glass surface and leave behind a thin line of fluid. But if $\Delta\sigma < 2\sigma_{water}$, a drop will leave behind a dry surface. Since raindrops usually leave behind a water tail, the minimum size of a drop that begins to slide down glass can be estimated as follows:

$$R_{min} = \sqrt{\frac{2\sigma_{water}}{\rho_{water}g}} \approx 2\,\mathrm{mm}\,(0.0787402\,\mathrm{in}). \qquad (7.48)$$

This estimation corresponds well with everyday observations.

Bubbles in Puddles When it Rains When it rains, we can sometimes see a considerable amount of bubbles form on the surface of puddles (Fig. 7.25). If it is raining hard, it seems that the water in puddles is boiling. When does this happen? It is hardly the case that every shower causes bubbles to appear in puddles. While learning about waterfalls, we analyzed the topic of drops that fall on water. When they fall, bubbles do not usually form. The vast majority of time, a slight depression forms in the water. When it straightens out, it gives rise to a water column. But if rain has a large amount of kinetic energy, i.e., the drops that fall are quite large, then a fairly deep channel will appear in the water column. This water channel will not straighten itself out but instead will collapse, thus turning into an underwater bubble, which will float up and appear on the puddle's surface.

Let's assume that the radius of a falling drop of water is R, the radius of a bubble that formed is R_0, and the length of the water channel that formed in the puddle is L. The radius of the channel is approximately equal to the radius of the drop and the air pressure in the channel and the bubble is equal to the pressure in the atmosphere. If we consider that the bubble is shaped like a hemisphere, we can make the air volume in the channel and the bubble equal: $\pi R^2 L \approx \frac{2}{3}\pi R_0^3$.

From this it follows that:

$$R \approx \sqrt{\frac{2R_0^3}{3L}}. \qquad (7.49)$$

Fig. 7.25   Bubbles in puddles when it rains

It stands to reason that a fairly deep channel cannot form in a shallow puddle. Thus, a puddle must be deep enough in order for bubbles to appear.

We will assume that $L \approx 1\,\text{cm}\,(0.393701\,\text{in})$ and the size of a bubble is $R_0 \approx 5\,\text{mm}\,(0.19685\,\text{in})$. The radius of a falling raindrop should then not be less than 2 mm (0.0787402 in). This is a very large drop. Hence, in order for us to see bubbles in puddles, two conditions must be met: raindrops should be large and puddles should be deep.

Another conclusion we can make when we see bubbles in puddles is that bubbles "live" for a long time only when the humidity level is close to 100%. There is even a saying that "bubbles on puddles mean rain for the next three days."[1] The reason for this is that bubbles disappear only when the liquid in their walls actively evaporates. If the humidity level is low, intensive evaporation occurs and bubbles quickly burst.

7.4   Atmospheric Electricity. Thunder and Lightning

Thunder, lightning, polar lights and St. Elmo's fire are all striking displays of atmospheric electricity.

The primary cause of these phenomena is the presence of ions—particles of molecular size that carry positive or negative electric charges—in the atmosphere. Ions form in the atmosphere due to the ionization of gas molecules

---

[1] This is the American equivalent of the Russian saying (translator's note).

that make up this process. The reason for ionization is, therefore, cosmic radiation and a steady flow of particles that come from space, primarily from the Sun. This flow penetrates through the atmosphere, the World Ocean and the Earth's crust. Ionization is the process of transferring enough energy to an atom so that a valence electron can be detached from it. As a result, the atom becomes a positively charged particle with the charge $e = 1.6 \times 10^{-19}$ C. The electron that was almost immediately released joins another atom and forms a negatively charged ion with the charge $- e$. The intensity of ion formation at the Earth's surface is on average $10^7 \, \frac{\text{ion pairs}}{\text{m}^3 \, \text{s}}$. The ions that form first usually bond with several molecules and with larger particles in the atmosphere, which creates stable charge complexes. Thus, there are ions of different sizes and masses with the charge $\pm e$ in the atmosphere.

The atmosphere is able to conduct electricity because ions exist there. If there is an electric field with the strength $\mathbf{E}$, an electric current is produced with a charge density that is equal to $j = \frac{E}{\rho}$ where ρ is the electrical resistivity of the atmosphere. The current strength that flows in the Earth's atmosphere and is directed toward each square meter of it is small—approximately 3×10^{-12} A. The density of atmospheric electricity $j = 3 \times 10^{-12}$ A/m^2 practically does not change with altitude because electric charges do not accumulate or disappear in the atmosphere. This means that a positive electric charge 3×10^{-12} C flows to one square meter of the Earth's surface within a second. The current strength that flows from the ionosphere to the Earth's surface is equal to: $I = 4\pi R_E^2 j = 1800$ A. The Earth's total negative charge is $q = 10^5$ C; the ionosphere is the same in magnitude but has a positive charge. The potential difference between the boundary of the ionosphere and the Earth's surface is 4×10^5 V (Fig. 7.26). The strength of the electric field near the Earth's surface has the highest value—about 100 V/m.

Thus, the electrical structure of the Earth resembles a giant round capacitor with the ionosphere's lower boundary and the Earth's surface (Fig. 7.27) carrying out the function of the capacitator's plates. This "earthly capacitor" runs out of power when the sky is clear and charges up when there is inclement weather such as rain and thunderstorms.

Cloud electrification is another phenomenon that helps understand the effects of atmospheric electricity. Let's analyze a drop of water that moves from the top to the bottom of a cloud due to gravity (Fig. 7.28).

Polarization of drops takes place in the Earth's electric field; moreover, their upper portion acquires a negative charge, while their lower portion is positive. As drops move downward, they begin to collide with positively and negatively charged ions. Negatively charged ions will be drawn to the bottom of the drops, while the positively charged ones, on the other hand, will be

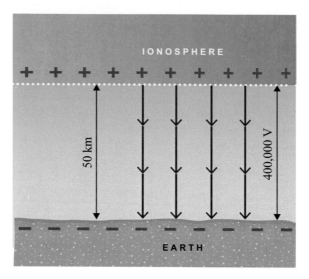

Fig. 7.26 The direction of an electric current between the ionosphere's lower boundary and the Earth's surface

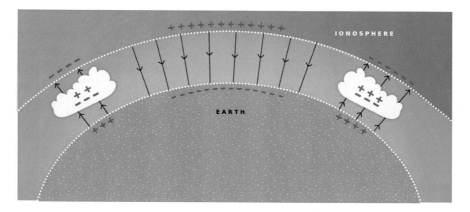

Fig. 7.27 Diagram of the electric charge distribution between the ionosphere, clouds and the Earth's surface, as well as the direction of an electric current

repelled by them. Consequently, when moving down the cloud, drops will acquire an even stronger negative charge. Since there are many drops, negatively charged ions will accumulate in the base of the cloud. Positively charged ions, which the drops repel, will be carried to the top of the cloud (Fig. 7.29) by convective currents of rising air.

If there are areas in the cloud where there is a large amount of electric charge, it becomes a storm cloud. Storm clouds tend to tower in size and may be up to several kilometers (feet) high. The area in a storm cloud with

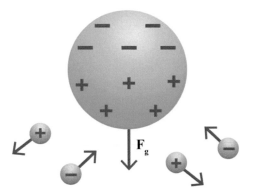

Fig. 7.28 Drops in a cloud that move downward due to gravity

Fig. 7.29 Distribution of charges in a cloud

a negative charge is normally located at an altitude of 2–3 km (6.56168–9.84252 ft) from the Earth's surface, while the positively charged area is 4–6 km (13.1234–19.685 ft) from it. These areas tend to be about 1 km (3280.84 ft) in either direction. The average electric charge of a storm cloud is ± 25 C for the area where there is a positive and a negative charge, respectively. The strength of an electric field between the base of a storm cloud and the Earth's surface reaches 10^4 V/m, while the strength of the field in the main charging area is 10^5 V/m.

A thunderstorm is an atmospheric phenomenon in which electric discharges, which appear as lightning, occur either between clouds and the Earth's surface or only between clouds.

> Cloud electrification may also occur in other circumstances, for example, when subcooled drops freeze, drops on ice particles collide and splash and ice crystals collide with one another.

What is lightning and how does it develop?

> Lightning is an electric discharge in the Earth's atmosphere that usually happens during a thunderstorm and is followed by a bright flash of light and a crack of thunder.

Cloud electrification is the driving force behind the formation of lightning. The fact that the base of a cloud carries a negative charge, while the top carries a positive one makes cloud-to-ground lightning similar to a gigantic capacitor. The lightning that we usually see during a thunderstorm is forked lightning. These are spark discharges approximately 15–20 km (49.2126–65.6168 ft) in length that branch out when they are close to the Earth's surface (Fig. 7.30). Lightning is several dozen centimeters (feet) in diameter and develops between the Earth's surface and clouds, between clouds and also inside of clouds. Lighting usually transfers a negative charge to the Earth.

There are three main stages of lightning development. The initial stage, which is called the *stepped leader*, lasts approximately 10^{-3} s. Free electrons, which are located in the lower edge of a storm cloud in a field with an intensity up to 10^6 V/m (621.371192 mi), significantly speed up. Because of a large amount of kinetic energy in this area, these electrons ionize the atoms and molecules with which they come in contact. New secondary electrons emerge, which also ionize atoms and so on. Thus, an abundance of high-speed electrons appears. Consequently, a conductive high-temperature channel (called a *stepped leader*) quickly arises (Fig. 7.31a). An electric charge is distributed along the leader channel, which reaches a magnitude of 4–5 C. The leader channel carries a current of hundreds of amperes charged with tens of megavolts of electricity. The leader head moves to the Earth's surface (Fig. 7.31b) at a tremendous speed—up to 10^7 m/s.

Fig. 7.30 Cloud-to-ground lightning

When the leader channel reaches the Earth's surface (Fig. 7.31c), the second and most important stage of the entire process begins: the main electric charge is transferred to the Earth. This significant stage is characterized by a sharp increase in the brightness of the channel's glow, loud noise (i.e., thunder) and the flow of the current impulse along the channel up to 200–300 kA in the span of about 10^{-3} s. In addition, roughly 10^9 J of energy is released and the temperature of the ionized gas in the channel climbs to 1000 K. Thunder is caused by a sound wave generated by a sudden increase in air volume due to a spike in temperature in the lightning channel. We hear cracks of thunder come from a flash of lightning after a certain period of time because of the fact that speed of sound in the air is about 330 m (1082.68 ft)/s. This is much slower than the speed of light, which is 3×10^8 m/s.

The length of the delay between a streak of lightning and a peal thunder is determined by the distance from an observer to the storm's epicenter.

In the last and final stage of lightning's development, there is a charge transfer through the lightning channel (Fig. 7.31d). This stage is followed by lower current values—up to 1 kA—and lasts about 1 ms.

An unusual type of lightning is ball lightning (Fig. 7.32). It is a luminous ball that usually appears after forked lightning occurs.

Ball lightning lasts from several seconds to several minutes, can strike inside of buildings and moves silently and slowly, sometimes exploding with

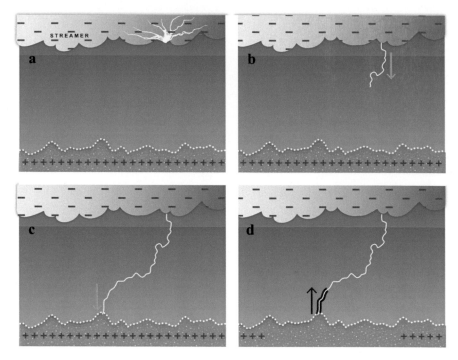

Fig. 7.31 The stages of lightning's development: **a** a stepped leader develops; **b** this stepped leader begins to move toward the Earth's surface; **c** the electric current closes (i.e., the stepped leader reaches the Earth's surface); **d** an electric current impulse along the lightning channel is set in motion

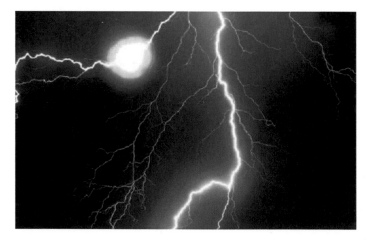

Fig. 7.32 Ball lightning

a loud pop. According to research done on ball lightning, it is a plasmoid, i.e., a gas consisting of ionized particles. We will estimate the energy of ball lightning with the radius $r = 10\,\text{cm}$ (3.93701 in), assuming the mass density is approximately equal to the air density $\rho = 1.2\,\text{kg/m}^3$. The latter assumption is based on the fact that onlookers usually see ball lightning floating in the air. The ionization energy is roughly equal to $E_{\text{ion}} = 1.3 \times 10^{-18}\,\text{J}$. We have:

$$E = \frac{4}{3}\pi r^3 \rho E_{\text{ion}} = 5 \times 10^3\,\text{J}. \tag{7.50}$$

To put this in perspective, this energy is roughly equal to the energy of two bullets shot from an AK-47 assault rifle.

Lightning Energy Let's measure the amount of energy in lightning. We will assume that the horizontal dimensions of storm clouds are $4 \times 4\,\text{km}$, the altitude from the clouds' base to the Earth's surface is 1 km (0.621371 mi), and the potential difference between the clouds and the Earth is $U = 10^9\,\text{V}$. We will also assume that the lightning energy is equal to the energy of a parallel-plate capacitor with sheets that have the surface $S = 1.6 \times 10^7\,\text{m}^2$ and a distance between the sheets $d = 10^3\,\text{m}$. The electrical capacity of this type of capacitor is equal to:

$$C = \frac{\varepsilon_0 S}{d} \quad \text{where } \varepsilon_0 = 8.8 \times 10^{-12}\,\frac{\Phi}{\text{m}}. \tag{7.51}$$

The energy of this capacitator is equal to:

$$E = \frac{1}{2}CU^2 = 7 \times 10^{10}\,\text{J}. \tag{7.52}$$

We deduced the upper bound of lightning energy because in reality not nearly all of the energy that is stored in a storm cloud is transferred to the Earth through a lightning strike. This amount of energy is enough, for example, for a large house or an apartment to function normally (i.e., operate with a heating system, electricity, etc.) for a month.

7.5 Types and Shapes of Snowflakes. Blizzards

Just like rain, snow forms in clouds from drops of water. The temperature in clouds and in the air near the Earth's surface determine whether rain or snow will fall from a storm cloud. If the temperature in a cloud is higher than the freezing point of water (0 °C, 32 °F), then, as we already know, the coalescence of drops results in them increasing in size and finally spilling out of the clouds as precipitation—in this case, rain. However, if the temperature is below freezing, droplets crystallize, freeze and turn into ice. But some of the ice particles stay in the clouds: the smallest ones that are less than 10^{-6} m remain and chaotically move around because they collide with molecules in the atmosphere. If, however, a microscopic drop of water falls on a dust particle and freezes, it can acquire enough mass to begin to move downward. A small ice crystal appears, the size of which does not initially exceed 0.1 mm (0.00393701 in) in diameter. When an ice particle moves downward in a cloud, it "gathers" drops of water along the way that freeze and increase the size and mass of this "first snowflake." But even such a fully formed ice particle as this will not automatically turn into snow because if the temperature near the Earth's surface is warm enough, snowflakes melt and become nothing more than rain.

Water crystals (i.e., ice) always develop by forming shapes with six-fold symmetry. The reason for this is because of water molecules' structure (see Fig. 6.23). Only 60 and 120° angles are possible between the branches of an ice crystal that develops in a cloud. Therefore, when looking at these ice crystals in a plane, they usually appear to be regular hexagons (Fig. 7.33).

New droplets (Fig. 7.34) gravitate toward the apexes of the hexagon where they freeze. This is exactly how so many different and beautiful shapes of snowflake stars form.

But snowflakes are always perfectly shaped stars only in stories about Santa Claus. In actuality, their shape (Fig. 7.35) is determined by the environment

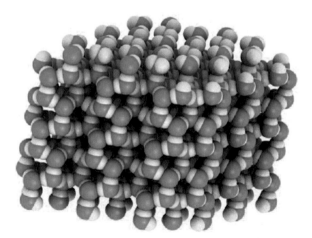

Fig. 7.33 The crystallized structure of ice with water molecules joined in regular hexagons (oxygen atoms are light blue and hydrogen atoms are white)

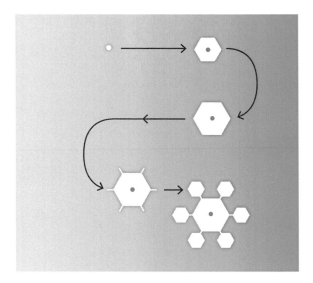

Fig. 7.34 Stages of a snowflake's development

in which they form. Scientists have counted nine major shapes of snow crystals (Fig. 7.36). In addition, each of these nine shapes further divides into other types and for this reason, 48 varieties of snowflakes exist.

The main reason for the difference in snowflakes' shapes is the temperature at which they form. If the temperature in a cloud is in the range of − 3 to 0 °C (27–32 °F), snowflakes develop that look like flat hexagonal plates. When the temperature is between − 5 and − 3 °C (23 and 27 °F), snowflakes

Fig. 7.35 An image of a snowflake taken under an electronic microscope (enlarged 36,000 times)

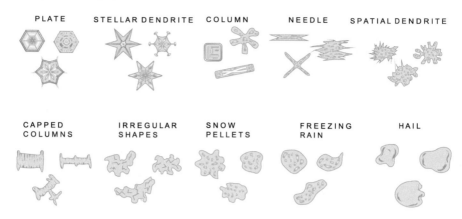

Fig. 7.36 Shapes of snow crystals

resemble needle crystals, but when it is between − 8 and − 5 °C (18 and 23 °F), they look like slender columns. When the temperature ranges from − 12 to − 8 °C (10–18 °F) snowflakes again look like flat hexagonal plates, but when it is between − 16 and − 12 °C (3 and 10 °F) the first snowflake stars appear. Lastly, at temperatures below − 35 °C (− 31 °F), small crystal prisms form in cirrus clouds.

In addition, the shape of snowflakes can change depending on their trajectory while falling. Snowflakes fall slowly; the speed at which they fall does not usually exceed 0.9 km (0.559234 mi)/h. This is because they consist of 95% air, which is the reason their density is low (100–400 kg/m³). If a snowflake spins as it is falling, its shape will be symmetrical, but if it falls a different way, for example, with one side sticking out in front, it will have an irregular shape. Falling snowflakes sometimes stick together and in this way large snowflakes form. Every large snowflake can have up to 200 snow crystals. Hence, a snowflake's shape gives us a clear understanding of the clouds' temperature on which it "traveled."

If the temperature of a cumulus cloud is close to 0 °C (32 °F), snow can become graupel or hail when there is extreme moisture. These types of precipitation are unique in that they do not have a pronounced crystal structure because the ice crystals in them develop rapidly and chaotically. Graupel consists of opaque pellets that are round or shaped like a cone and range in size from 1 to 15 mm (0.0393701–0.590551 in). When the weather is warm, we can sometimes see another form of precipitation such as hail. Hailstones are lumps of snow covered with a crust of ice. This crust forms when snowflakes that have stuck together in little clumps move inside of a cloud that has a high amount of supercooled drops. Drops of water stick to these little clumps and freeze, which causes sleet to form. If a cloud is very thick, this process can repeat itself over and over in which case hailstones become large and multilayered.

The size of a hailstone depends on the altitude at which it formed and how long it was in a cloud. Hailstones can range in size from fractions of a millimeter (less than one inch) to several centimeters (inches). Since hailstones fall much faster than snowflakes (sometimes faster than 15 m [49.2126 ft]/s) due to their large mass, even hailstones and ice particles do not have a chance to melt in summer despite the high atmospheric temperature near the Earth's surface.

Why does snow appear white to us? Ice crystals are, after all, transparent. Since each snowflake is an ice crystal, when isolated from other snowflakes, each one is transparent. This is because light passes through them without being reflected. But when there are a large number of snowflakes, the light from each and every wavelength is reflected off the surface of snowflakes of different shapes and sizes. We perceive this reflected light as the color white.

A blizzard is caused by a snowstorm mixed with wind, but it can even occur when there isn't a snowstorm because the wind may pick up and blow the snowflakes around that make up a blanket of snow. When this happens, strong winds blow snow so that it forms a thin layer (20–30 cm high [7.87402–11.811 in]) above the ground. We can analyze the development of a ground blizzard in a more simplistic manner by using the following method of reasoning. Let's consider a strip of a snowfield with the width D that runs in the direction that the wind is blowing (Fig. 7.37).

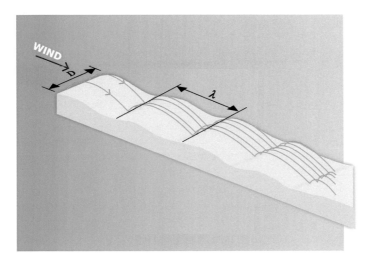

Fig. 7.37 An illustration showing how a blizzard forms

Fig. 7.38 Snow waves

First, it is important to note that the terrain of this snowfield is covered with snow waves (Fig. 7.38).

This is due to the fact that snowflakes' movement resembles jumps of a certain length λ (Fig. 7.37), which is determined by wind strength. The length λ varies anywhere from 1 m (3.28084 ft) for a gentle breeze to dozens of meters (hundreds of feet) for a strong wind. This wavy structure, which first appears in the nucleation center, further develops because of an eddy current from the windward side of the snow wave (Fig. 7.39). When snow is under pressure from wind, it gets carried from the windward to the leeward side of the waves. Snow waves move in a specific way, which obviously stops when the wind is no longer blowing.

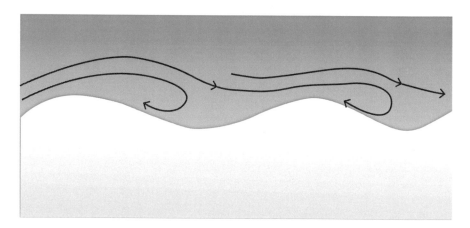

Fig. 7.39 The direction of airflow over snow waves

Fig. 7.40 Barkhan dunes and waves on sand

Thus, the wind gets ahold of snowflakes on the windward side of snow waves and carries them an average distance λ. A snowflake falls on the slope of a neighboring wave and forces one or more snowflakes out of it. If the snowflake that has fallen on the slope forces out $n > 1$ snowflakes, a blizzard intensifies and pulls in more and more of the snow drift. Let us assume that N snowflakes are in the air in a unit of time at the front of the snowy strip on which we are focusing. After k consecutive jumps at the distance $L = k\lambda$ from the onset of motion, Nn^k snowflakes will have flown up in the air. Consequently, the amount of snow that has been added to the mass that was set in motion will be the exponential function of length L, $m_{\text{snow}} \sim n^{\frac{\lambda}{L}}$.

Research done on blizzards has shown that usually $n = 1.1$; in other words, out of ten snowflakes only one of them forces out two new ones, but nine snowflakes each force out one new one.

If the wind that is carrying the snow encounters an obstacle—a tree, local topography, etc.—its speed decreases and a snowdrift forms.

For the same reason that waves appear on snow, barkhan dunes and sand hills can also be found on sand (Fig. 7.40).

7.6 Snow Avalanches in the Mountains

It is well known how dangerous snow avalanches are in the mountains (Fig. 7.41). There are several reasons why they occur, the first of which is that because of heavy snowfalls, snow accumulates on the leeward slope of

Fig. 7.41 An avalanche in the Caucasus Mountains

a mountain. The second reason is snow cover on the slope melts when it thaws. Third and lastly, an avalanche can be triggered by an environmental factor such as strong air concussion or an earthquake.

Before analyzing the causes of avalanches in the mountains, let's consider why snow tends to accumulate on the leeward slope, which makes these areas especially dangerous during a heavy snowfall. Let us assume that there is a snow-producing cloud over a mountain and the wind is blowing along the mountain's surface (Fig. 7.42).

Let us also assume that $\mathbf{v}_s$ is the vertical intensity of the speed of the snowflakes that are falling on the slope and v is the speed of the wind that is blowing parallel to the mountain's surface. The condition $v \sin \alpha < \mathbf{v}_s$ has been satisfied at the base of the mountain (Zone 1). Therefore, snow will fall and cover it. As soon as the slope becomes steeper (Zone 2), the wind will change direction and go around the slope at a sharper angle, which means that $v \sin \beta < \mathbf{v}_s$. In this spot, the snow will fly up without forming a snow cover. Snowflakes will settle on the mountain's surface close to the peak (Zone 3), but it is highly likely that they will be carried away by the wind. Turbulent flow patterns will then eventually begin on the leeward slope (Zone 4), which will help settle down the bulk of the snow. It is in this very spot that a snow avalanche is most likely to occur.

Now we will determine when a mass of snow that has accumulated on a slope may begin to move. We will focus on a specific amount with the mass m, which is lying on a mountain slope at the angle α (Fig. 7.43).

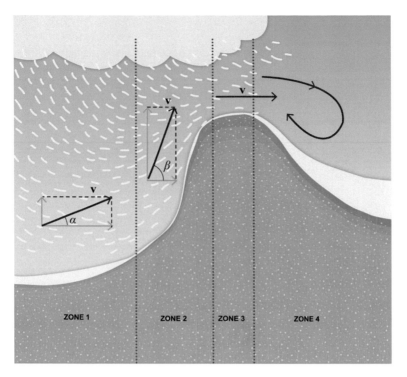

Fig. 7.42 Wind direction close to a mountain and to an area where there is abundant snow accumulation

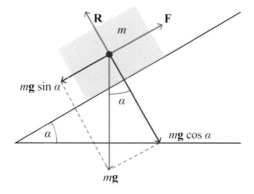

Fig. 7.43 The forces that act on a snowy mass on a slope

The force of gravity mg, the friction force F_{frict} and the reaction force are applied to the center of this mass from the side of the slope R. Newton's second law regarding projection of force provides two equations:

$$mg \cos \alpha = R; \qquad (7.53)$$

$$mg \sin \alpha - F_{\text{frict}} = ma. \qquad (7.54)$$

Here a is the acceleration that is channeled along the slope.

Which factors determine the magnitude of the friction force F_{frict}? It is first determined by friction of rest, which is similar to the friction that is found between two solid bodies lying one on top of the other. Specific details related to the fact that both the body lying on top and the slope itself are made up of snow particles and ice are also added to this scenario. It stands to reason that the bottom layer of snow doesn't fall off the slope because it has frozen in place and, in addition, the snow on top will also keep the mass we are analyzing from falling thanks to cohesive forces between the snowflakes. In order to understand what can trigger an avalanche, it should first of all be pointed out that F_{frict} always has the highest and most extreme value in this situation. We will denote it as F_{max}. This value is determined by many factors: the tilting angle, weather conditions (i.e., primarily temperature and humidity), the slope surface structure and the microstructure of the snow lying on the slope.

The condition of restricted movement of the mass of snow on the slope is $mg \sin \alpha = F_{\text{frict}}$. As soon as this constraint is broken ($mg \sin \alpha > F_{\text{max}}$), the mass of snow begins to accelerate uniformly down the slope.

What are some possible scenarios as to what it will end up doing? The first possibility is that this mass m will increase in size because of the snowfall. If it becomes this large, then

$$m > \frac{F_{\text{max}}}{g \sin \alpha}, \qquad (7.55)$$

the avalanche will begin to move at the accelerated rate

$$a = g \sin \alpha - \frac{F_{\text{max}}}{M}. \qquad (7.56)$$

A third variant is also possible, and it is that the snowstorm stops and the mass of snow on the slope does not change, but the environmental conditions do. The temperature, for example, may rise or fall. In this situation, the circumstances regarding the likelihood that snow will stick to the slope can change, which will lead to a decrease in the limiting value of the adhesive force F_{max}. If this occurs, the equilibrium condition may again be broken, which will cause an avalanche. Still another scenario may play out. If the avalanche on the slope is "hanging by a thread" and the equilibrium condition can be broken at any moment, even a minor external influence, for example, a gunshot, a shout or even hand clapping, may be enough to trigger it.

The sound waves generated in this case provide an additional amount of force that pushes an avalanche downhill.

Avalanches may be set off by a very small snow mass that has begun to slide down a slope and after a period of time they can pull in up to 1 hm^3 of snow and reach speeds of up to 50 m (164.042 ft)/s.

7.7   Reasons for Climate Change on the Earth

We have all heard of ice ages—those periods when the Earth's surface temperature dropped significantly. Studies of glaciers have provided evidence that ice ages existed in our planet's history. As we have already mentioned, before a glacier melts, it deposits a huge mass of boulders that it has transported from mountains. The areas where this debris was deposited are called *terminal moraines*. One can determine the geographic scale that glaciers covered by focusing specifically on the location of terminal moraines. It was found that about 125,000 years ago a glacier that originated in Scandinavia reached the latitude of Moscow. Moreover, approximately 250,000 years ago another glacier went as far south as the latitude of the Russian city Rostov-on-Don. Similar contour patterns of terminal moraines have been discovered in Europe and North America. Studies have shown that there were several dozen ice ages (Fig. 7.44) in the history of the Earth. They repeated at different intervals: from 40,000 to 350,000–400,000 years. The last ice age occurred roughly 15,000–20,000 years ago. At that time, all of the Earth's surface was covered with glacier ice and the overall number of glacial masses on it doubled (Fig. 7.45). The water that formed all of this ice came from the World Ocean, which subsequently resulted in a lowering of its level by approximately 100 m (328.084 ft). Between periods of the ice ages, the average temperature on the Earth rose and almost reached its average temperature today.

By looking at Fig. 7.46 it is possible to see how the average temperature on the Earth's surface changed over the course of the last 400,000 years.

Let us consider the main reasons for these changes in the Earth's climate. When we were examining the Earth's atmosphere, we maintained that the average temperature of a planet is determined by the balance between the energy that comes from the Sun and the energy that heats the Earth's surface without being radiated back into space.

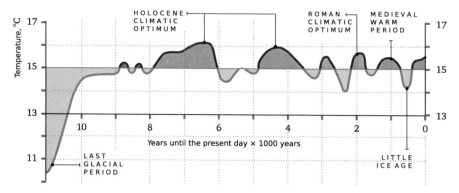

Fig. 7.44 Fluctuations in the Earth's average temperature after the Last Glacial Period

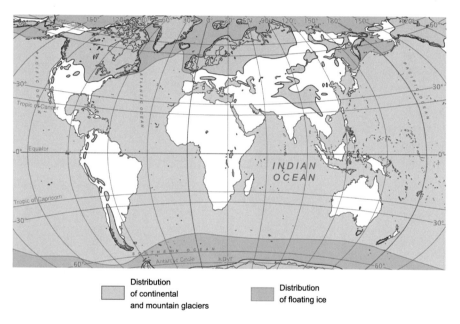

Fig. 7.45 Areas to which glaciers and drifting ice spread during the Last Glacial Maximum

The effective radiating temperature of the Earth's surface is its radiative equilibrium temperature.

This temperature is approximately equal to 249 K = − 24 °C (− 11.2 °F).

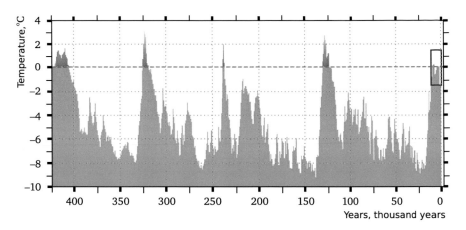

Fig. 7.46 Changes in the Earth's average temperature over the past 400,000 years (the area in the red box corresponds to the time interval shown in Fig. 7.44) in relation to today's readings

Effective Radiating Temperature of the Earth's Surface The luminosity that the Sun gives our planet is equal to $P = L\dfrac{R_E^2}{4R_{E-S}^2}$, where $L = 3.8 \times 10^{27}\,\text{V}$ is the Sun's luminosity. If $A = 0.28$ is the Earth's albedo, then the amount of energy that heats the Earth is $P(1 - A)$. The total energy that is emitted by the Earth is $kT_{rad}^4 \times 4\pi R_E^2$, where $k = 1.38 \times 10^{-23}\,\text{J/K}$ is the Boltzmann constant and T_{rad} is the effective radiating temperature of the Earth. We will make the total solar radiation equal to the Earth's total infrared radiation. We get

$$kT_{rad}^4 \times 4\pi R_E^2 = (1 - A)L\frac{R_E^2}{4R_{E-S}^2} = (1 - A)kT_S^4\frac{R_E^2}{4R_{E-S}^2} \qquad (7.57)$$

(T_S is the temperature on the Sun's surface). From this it follows that:

$$T_{\text{rad}} = T_S \sqrt[4]{1 - A}\sqrt{\frac{R_E}{2R_{E-S}}} = 255\,\text{K}. \qquad (7.58)$$

We know that due to the greenhouse effect and the transfer of energy from the oceans to the atmosphere, the actual temperature of the Earth's surface is much higher than its effective radiating temperature.

Important factors that can significantly change over time in this formula are the Earth's albedo and the average distance from the Earth to the Sun. The greater the amount of ice and snow on the Earth's surface, the higher albedo is because ice and snow reflect solar radiation very well. But solar radiation's dependence on albedo is quite weak: a 10% change in albedo only results in a 1% change of T_{rad}, which is about 3°. As far as the distance from the Earth to

Sun is concerned, it can change because of astronomical phenomena, which we examined in Chap. 1.

What then are the astronomical factors that affect the change in illumination of the Earth's hemispheres (Fig. 7.47)? The first factor is the precession of the Earth's axis and its rotation with a period of about 25,700 years because of the Sun and the Moon's influence on it.

The second factor is variations in the tilting angle of the Earth's axis toward the plane of its orbit caused by the influence exerted on it by other planets in the Solar System. The period of these variations is roughly 41,000 years. Finally, the third factor is variations in the eccentricity of the Earth with a period of about 93,000 years. Each of these factors leads to periodic changes in the Earth's illumination.

Since each of these circumstances periodically repeat themselves—although their periods are different—long intervals of time occur when certain factors may intensify or, conversely, reduce the effect of other ones. As a result, the Earth's climate experiences warm and cold periods when the temperature significantly differs from average readings. Small percentage changes in average illumination are enough for the Earth to warm up or cool down a few degrees. And that is already a major climate change! Figure 7.48 illustrates how the simultaneous influence of several astronomical factors results in alternating warm and cold periods on the Earth.

Let's qualitatively explore the physical processes that explain why the Earth's surface warms up and cools down. We have already thoroughly considered the axial precession of the Earth's axis and we are aware that the main reason it occurs is because of the mutual influence between the Earth and the Moon. However, precession is not only caused by the Moon, but also by the Sun, and, to a lesser extent, other planets contribute to it as well. An

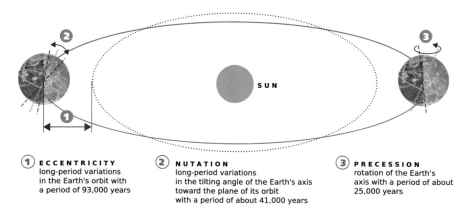

① ECCENTRICITY
long-period variations
in the Earth's orbit with
a period of 93,000 years

② NUTATION
long-period variations
in the tilting angle of the Earth's axis
toward the plane of its orbit
with a period of about 41,000 years

③ PRECESSION
rotation of the Earth's
axis with a period of about
25,000 years

Fig. 7.47 Astronomical factors that affect the Earth's climate

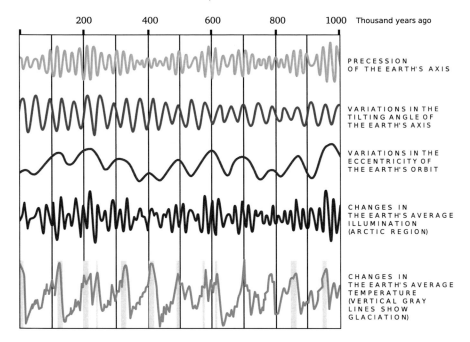

Fig. 7.48 The impact of astronomical factors on the Earth's climate

analogous situation brings about a slow and almost periodic change in the inclination of the Earth's axis to the ecliptic. This is known as *nutation*.

The reason for these variations in the eccentricity of the Earth's orbit is because of the influence that the movement of Jupiter, Venus, and, to a lesser extent, other planets exert upon it. The mean eccentricity of the Earth's orbit is 0.028. It is now less than average—0.017—and continues to decrease. In 25,000 years, the Earth's orbit will be almost circular and the maximum value of eccentricity will be 0.0658, which is four times higher than it is at present. Although the magnitude of eccentricity is small, its variations have a significant impact upon the Earth's climate. Estimates suggest that the dependence of the Earth's integral illumination on eccentricity is as follows: $E \sim \left(1 - e^2\right)^{-\frac{1}{2}} \approx 1 + \frac{1}{2}e$. In other words, the larger the eccentricity, the greater the Earth's illumination. However, the magnitude of eccentricity is much less than unity. This means that a change in eccentricity exerts a minimal influence on illumination. Even when eccentricity is at its maximum value, the energy that the Earth receives per year increases by only 0.2%. But if we calculate the average temperature change on the Earth, then this adjustment will require that we find the fourth root, which will further reduce the effect fourfold. However, we must take the information that follows into consideration.

We have come to find out that in order to correctly assess the influence that astronomical factors have on the climate, one must account for the integral energy that the Earth receives during certain seasons of the year, specifically, winter and summer. Given the disbalance of land distribution in the Northern and Southern Hemispheres of the Earth (i.e., there is much more land in the Northern Hemisphere), one is able to calculate the warming and cooling periods, both in the past and in the future, fairly accurately.

Now we will measure how the Earth's illumination changes during the course of a year as a result of orbital ellipticity. The distance from the Earth to the Sun is maximum at aphelion and depends on an eccentricity such as $1+e$. However, illumination is inversely proportional to squared distance. Therefore, its dependence on e is as follows: $(1 + e)^{-2} \approx 1 - 2e$. At perihelion, the distance is, by contrast, minimal and proportional to $1 - e$, but illumination is minimal and proportional to $(1 - e)^{-2} \approx 1 + 2e$. Hence, a difference in illumination at perihelion and aphelion such as $4e$ is considered average; in other words, when eccentricity is at its maximum, this difference can reach 26%, which is a significant amount. Two factors must be met in order for the average temperature on the Earth's surface to drop and thus cause an ice age. The first is that the Earth's orbit must have very high eccentricity. The second is that the date when the Earth reaches its perihelion must be close to the date when the winter solstice occurs in the Northern Hemisphere.

Let's imagine what happens to the Earth's climate system when these astronomical parameters come together. It is summer in the Northern Hemisphere. The Earth is located in a remote place in orbit and is slowly moving in it. Thus, summer in the Northern Hemisphere becomes cool and lasts longer. Glaciers in the Northern Hemisphere melt more slowly than usual. These are exactly the conditions that can bring about the onset of glaciation. At this time in the Southern Hemisphere a long and cold winter is underway. The seasons change. In the Northern Hemisphere, winter sets in. The Sun is close to the Earth, but cannot melt the snow and ice. In the Southern Hemisphere a hot summer is underway, but only the ocean becomes warmer because there are practically no glaciers in the Southern Hemisphere except in Antarctica. In conditions such as these, the total area on the Earth covered by glaciers will, in essence, dramatically increase.

Eleven thousand years pass and perihelion now occurs at the same time as the summer solstice. Within that timeframe, eccentricity has not changed significantly, but has stayed at a high level. Now there is a long, severe winter in the Northern Hemisphere. It gives way, however, to a hot summer and the glaciers in the Northern Hemisphere have time to melt to a significant extent. In the Southern Hemisphere, on the other hand, conditions are presently developing that are conducive to glacier growth, but there is not much land available there where they can form. Consequently, the total area that glaciers occupy is decreasing. The Earth's albedo is becoming smaller and the Earth is beginning to absorb more solar radiation. This is causing a rise in the average temperature on our planet. Thus, an ice age has given way to a warming period (see Fig. 7.46).

The Characteristic Time of Climatic Fluctuations We will measure the characteristic time of climatic fluctuations. The perihelion of the Earth's orbit moves relative to the stars and makes a complete orbital revolution within $T_c = 100{,}000$ years. The winter and summer solstices, however, orbit the Earth in the opposite direction with a period that is equal to the precession period $T_{prec} = 26{,}000$ years. Based on this information, we can calculate the

Fig. 7.49 Illustration of the Earth's climate change

period T_0 when perihelion and the winter solstice repeatedly align:

$$\frac{1}{T_0} = \frac{1}{T_{\text{prec}}} + \frac{1}{T_{\text{c}}} \tag{7.59}$$

from which $T_0 = 21{,}000$ years. This is roughly the period when cold spells began on the Earth.

In which phase of this cycle are we now? The Earth is at its perihelion on January 4th. This is close to the winter solstice, which falls on December 22nd. Relatively recently—768 years ago—the winter solstice was on the same day as the Earth reached its perihelion. Prior to that, this had happened 22,000 years earlier. At that time, there was more eccentricity than there is now (0.020) and the Ice Age was truly wreaking havoc on the Earth. It was then that a glacier floated to the latitude of Moscow. You are probably wondering why we don't have an ice age now because, after all, perihelion occurs at practically the same time as the winter solstice. The answer is that the Earth's climate today is not at all like it was during the Ice Age and, what is more, the area covered by glaciers has decreased over the last 100 years. Within that time, the Earth's average temperature has increased by about 1 °C (34 °F), while the level of the World Ocean has risen by approximately 10 cm (3.93701 in). The simple truth is that today there is a small amount of eccentricity and, in addition, it continues to decrease. Because of this, our lifetime and the next millennium will be one of the warmest periods in the last million years (Fig. 7.49).

But what will happen to our planet's climate in the future?

The next time perihelion occurs at the same time as the winter solstice will not be soon; this will take place in almost 20,000 years. During that entire time, sunlight in the Northern Hemisphere will diminish. However, even a minimal amount of sunlight after 20,000 years will not cause significant glaciation because at that time eccentricity will be close to zero.

Everyone on the Earth can rest assured that our planet's climate is not in any danger, at least not for tens of thousands of years.

Further Reading

1 Blum, A.: The Weather Machine: A Journey Inside the Forecast. Ecco (2019)
2 Boerner, H.: Ball Lighting: A Popular Guide to a Longstanding Mystery in Atmospheric Electricity. Springer (2019)
3 Byalko, A.V.: Our Planet the Earth. MIR Publisher (1983)
4 Dessler, A.E.: Introduction to Modern Climate Change, 3rd edn. Cambridge University Press (2021)
5 Henning, R.: Field Guide to the Weather: Learn to Identify Clouds and Storms, Forecast the Weather, and Stay Safe. Adventure Publications (2019)
6 Libbrecht, K.G.: Snow Crystals: A Case Study in Spontaneous Structure Formation. Princeton University Press (2021)
7 Maslin, M.: Climate Change: A Very Short Introduction, 4th edn. Oxford University Press (2021)

8 Mason, B.J.: The Physics of Clouds, 2nd edn. Oxford University Press (2010)
9 Pudasaini, S.P., Hutter, K.: Avalanche Dynamics. Dynamics of Rapid Flows of Dense Granular Avalanches. Springer (2007)
10 Shonk, J.: Introducing Meteorology: A Guide to the Weather (Introducing Earth and Environmental Sciences). Dunedin Academic Press (2020)
11 Volland, H.: Atmospheric Electrodynamics. Springer-Verlag (1984)

Printed in the United States
by Baker & Taylor Publisher Services